MEMOIRS
of the
American Mathematical Society

Volume 232 • Number 1092 (fourth of 6 numbers) • November 2014

A Power Law of Order 1/4 for Critical Mean Field Swendsen-Wang Dynamics

Yun Long
Asaf Nachmias
Weiyang Ning
Yuval Peres

ISSN 0065-9266 (print)   ISSN 1947-6221 (online)

American Mathematical Society
Providence, Rhode Island

Library of Congress Cataloging-in-Publication Data

A power law of order 1/4 for critical mean field Swendsen-Wang dynamics / Yun Long [and three others]
    pages cm. – (Memoirs of the American Mathematical Society, ISSN 0065-9266 ; volume 232, number 1092)
    Includes bibliographical references.
    ISBN 978-1-4704-0910-4 (alk. paper)
    1. Markov processes.   2. Spin waves–Mathematical models.   I. Long, Yun, 1982-
QA274.7.P68   2014
519.2′33–dc23   2014024667

DOI: http://dx.doi.org/10.1090/memo/1092

Memoirs of the American Mathematical Society

This journal is devoted entirely to research in pure and applied mathematics.

Subscription information. Beginning with the January 2010 issue, *Memoirs* is accessible from www.ams.org/journals. The 2014 subscription begins with volume 227 and consists of six mailings, each containing one or more numbers. Subscription prices are as follows: for paper delivery, US$827 list, US$661.60 institutional member; for electronic delivery, US$728 list, US$582.40 institutional member. Upon request, subscribers to paper delivery of this journal are also entitled to receive electronic delivery. If ordering the paper version, add US$10 for delivery within the United States; US$69 for outside the United States. Subscription renewals are subject to late fees. See www.ams.org/help-faq for more journal subscription information. Each number may be ordered separately; *please specify number* when ordering an individual number.

Back number information. For back issues see www.ams.org/bookstore.

Subscriptions and orders should be addressed to the American Mathematical Society, P. O. Box 845904, Boston, MA 02284-5904 USA. *All orders must be accompanied by payment.* Other correspondence should be addressed to 201 Charles Street, Providence, RI 02904-2294 USA.

Copying and reprinting. Individual readers of this publication, and nonprofit libraries acting for them, are permitted to make fair use of the material, such as to copy select pages for use in teaching or research. Permission is granted to quote brief passages from this publication in reviews, provided the customary acknowledgment of the source is given.

Republication, systematic copying, or multiple reproduction of any material in this publication is permitted only under license from the American Mathematical Society. Permissions to reuse portions of AMS publication content are handled by Copyright Clearance Center's RightsLink® service. For more information, please visit: http://www.ams.org/rightslink.

Send requests for translation rights and licensed reprints to reprint-permission@ams.org.

Excluded from these provisions is material for which the author holds copyright. In such cases, requests for permission to reuse or reprint material should be addressed directly to the author(s). Copyright ownership is indicated on the copyright page, or on the lower right-hand corner of the first page of each article within proceedings volumes.

Memoirs of the American Mathematical Society (ISSN 0065-9266 (print); 1947-6221 (online)) is published bimonthly (each volume consisting usually of more than one number) by the American Mathematical Society at 201 Charles Street, Providence, RI 02904-2294 USA. Periodicals postage paid at Providence, RI. Postmaster: Send address changes to Memoirs, American Mathematical Society, 201 Charles Street, Providence, RI 02904-2294 USA.

© 2014 by the American Mathematical Society. All rights reserved.
This publication is indexed in *Mathematical Reviews*®, *Zentralblatt MATH*, *Science Citation Index*®, *Science Citation Index*^{TM}-*Expanded*, *ISI Alerting Services*^{SM}, *SciSearch*®, *Research Alert*®, *CompuMath Citation Index*®, *Current Contents*®/*Physical, Chemical & Earth Sciences*. This publication is archived in *Portico* and *CLOCKSS*.
Printed in the United States of America.

∞ The paper used in this book is acid-free and falls within the guidelines established to ensure permanence and durability.
Visit the AMS home page at http://www.ams.org/

10 9 8 7 6 5 4 3 2 1   19 18 17 16 15 14

Contents

Chapter 1. Introduction — 1

Chapter 2. Statement of the results — 3
  2.1. Random graph estimates — 4

Chapter 3. Mixing time preliminaries — 7

Chapter 4. Outline of the proof of Theorem 2.1 — 9
  4.1. Outline of the Proof of Theorem 2.1 (i) — 9
  4.2. Outline of the Proof of Theorem 2.1 (iii) — 10
  4.3. Outline of the Proof of Theorem 2.1 (ii) — 10

Chapter 5. Random graph estimates — 13
  5.1. The exploration process — 13
  5.2. Random graph lemmas for non-critical cases — 14
  5.3. Random graph lemmas for the near-critical case — 19

Chapter 6. Supercritical case — 37

Chapter 7. Subcritical case — 45

Chapter 8. Critical Case — 49
  8.1. Starting at the $[n^{3/4}, n]$ regime: Proof of Theorem 8.1 — 50
  8.2. Starting at the $[0, n^{3/4}]$ regime: Proof of Theorem 8.2 — 71
  8.3. The lower bound on the mixing time — 77

Chapter 9. Fast mixing of the Swendsen-Wang process on trees — 79

Acknowledgements — 81

Bibliography — 83

Abstract

The Swendsen-Wang dynamics is a Markov chain widely used by physicists to sample from the Boltzmann-Gibbs distribution of the Ising model. Cooper, Dyer, Frieze and Rue proved that on the complete graph K_n the mixing time of the chain is at most $O(\sqrt{n})$ for all non-critical temperatures. In this paper we show that the mixing time is $\Theta(1)$ in high temperatures, $\Theta(\log n)$ in low temperatures and $\Theta(n^{1/4})$ at criticality. We also provide an upper bound of $O(\log n)$ for Swendsen-Wang dynamics for the q-state ferromagnetic Potts model on any tree of n vertices.

Received by the editor July 14, 2011, and, in revised form, February 14, 2013.
Article electronically published on March 11, 2014.
DOI: http://dx.doi.org/10.1090/memo/1092
2010 *Mathematics Subject Classification.* Primary 60J10, 82B20.
Key words and phrases. Markov chains, Mixing time, Ising model, Swendsen-Wang algorithm

©2014 American Mathematical Society

CHAPTER 1

Introduction

Local Markov chains (e.g. "Glauber dynamics") are commonly used to sample from spin systems on graphs. At low temperatures, however, their mixing time becomes very large (sometimes exponential in the size of the graph), making it computationally harder to sample from the equilibrium measure. In some cases, "Global" Markov chains, which allow moves like cluster flipping, yield much faster mixing and those are the algorithms of choice when practitioners actually sample (see for example [31], [32], [33] and [36]; see [17] for a different polynomial time algorithm for sampling from the Ising model). The Swendsen-Wang (SW) algorithm for the q-state ferromagnetic Potts model and its variations are frequently used in practice. Gore and Jerrum [15] discovered that for any $q > 2$, on the complete graph K_n there are temperatures where the SW dynamics has mixing time of order at least $\exp(\Omega(\sqrt{n}))$. Borgs, Chayes, Frieze, Kim, Tetali, Vigoda and Vu [5] proved a similar lower bound on the mixing time of the SW algorithm at the critical temperatures, on the d-dimensional lattice torus for any $d \geq 2$ and q sufficiently large.

The natural question remaining is how does the SW algorithm perform when $q = 2$ (i.e., for the Ising model). The first positive result in this direction is due to Cooper, Dyer, Frieze and Rue [7]. They proved that the SW algorithm on the complete graph on n vertices has mixing time at most $O(\sqrt{n})$ for all non-critical temperatures. In this paper we show that the mixing time is $\Theta(1)$ in high temperatures, $\Theta(\log n)$ in low temperatures and $\Theta(n^{1/4})$ at the critical temperature. The study of the mixing time at criticality is the main effort of this paper. Heuristic arguments for the exponent $1/4$ at criticality were found earlier by physicists, see [30] and [27].

It is instructive to compare these results with the mixing time of Glauber dynamics for the critical Ising model on K_n. In [20], the authors show that this mixing time is $\Theta(n^{3/2})$. Since in the SW dynamics we update n vertices in each step, the number of vertex updates up to the mixing time is $\Theta(n^{5/4})$, that is, it performs faster by $n^{1/4}$ than Glauber dynamics.

CHAPTER 2

Statement of the results

The mixing time of a finite Markov chain with transition matrix **p** is defined by
$$T_{\mathrm{mix}} = T_{\mathrm{mix}}(1/4) = \min\left\{t : \|\mathbf{p}^t(x,\cdot) - \pi(\cdot)\|_{\mathrm{TV}} \leq 1/4 \,, \text{for all } x \in V\right\},$$
where $\|\mu - \nu\|_{\mathrm{TV}} = \max_{A \subset V} |\mu(A) - \nu(A)|$ is the total variation distance. Before describing the Swendsen-Wang (SW) algorithm on a graph $G = (V, \mathcal{E})$, let us first describe its stationary distribution, also known as the Ising model. This is a probability measure (also known as the Boltzmann-Gibbs distribution) on the set $\Omega = \{1, -1\}^V$ where the probability of each $\sigma \in \Omega$ is
$$\mathbf{P}(\sigma) = \frac{e^{\beta \sum_{(u,v) \in \mathcal{E}} \sigma(u)\sigma(v)}}{Z(G)},$$
where $\beta \in [0, \infty]$ is a parameter usually referred to as the *inverse temperature*, and the *partition* function $Z(G)$ is defined by
$$Z(G) = \sum_{\sigma \in \Omega} e^{\beta \sum_{(u,v) \in \mathcal{E}} \sigma(u)\sigma(v)}.$$
For $\sigma \subset \Omega$, let $G^+(\sigma)$ be the graph spanned by the vertices of G which are assigned 1 by σ and similarly let $G^-(\sigma)$ be the graph spanned by the vertices of G which are assigned -1 by σ. The SW dynamics on a graph G with percolation parameter $p \in [0, 1]$ is a Markov chain on Ω. Given the current state of the chain σ_t, we obtain the next state σ_{t+1} by the following two-step procedure:

1. Perform independent p-bond percolation on $G^+(\sigma_t)$ and on $G^-(\sigma_t)$ separately. That is, retain each edge of $G^+(\sigma_t)$ and $G^-(\sigma_t)$ with probability p and erase with probability $1 - p$, independently for all edges. Call the obtained graphs G_p^+ and G_p^-, respectively.
2. To obtain σ_{t+1}, for each connected component $\mathcal{C}$ of G_p^+ and of G_p^-, with probability $1/2$ assign all vertices of $\mathcal{C}$ the same sign 1 and with probability $1/2$ assign them all the sign -1, independently for all these components.

It is easy to show using Fortuin and Kasteleyn's Random Cluster model [14] (see Edwards and Sokal [11] for this derivation) that the Ising model measure is invariant under the SW dynamics when $p = 1 - e^{-2\beta}$. Moreover, the SW dynamics is clearly an aperiodic and irreducible Markov chain. Hence from any starting configuration σ_0, the law of σ_t obtained after t updates, converges in distribution to the stationary Ising measure. Cooper, Dyer, Frieze and Rue [7] investigated the mixing time of the SW dynamics on the complete graph on n vertices. They showed that if $p = \frac{c}{n}$ when $c \in (0, \infty) \setminus \{2\}$ is some constant independent of n, then the mixing time of the dynamics is at most $O(\sqrt{n})$. The following Theorem improves their result by giving the precise order of the mixing time at all temperatures.

THEOREM 2.1. *Consider the SW dynamics on the complete graph K_n on n vertices, with percolation parameter $p = \frac{c}{n}$, where c is a constant independent of n. Then,*

  (i) *If $c > 2$ then $T_{\mathrm{mix}} = \Theta(\log n)$.*
  (ii) *If $c = 2$ then $T_{\mathrm{mix}} = \Theta(n^{1/4})$.*
  (iii) *If $c < 2$ then $T_{\mathrm{mix}} = \Theta(1)$.*

The q-state ferromagnetic Potts model with parameter $\beta > 0$ on a graph $G = (V, \mathcal{E})$ is a probability measure on the set $\Omega = \{1, 2, \cdots, q\}^V$ where for each $\sigma \in \Omega$

$$\mathbf{P}(\sigma) = \frac{e^{\beta \sum_{(u,v) \in \mathcal{E}} \mathbf{1}_{\{\sigma(u)=\sigma(v)\}}}}{Z(G)},$$

and the *partition* function $Z(G)$ is defined by

$$Z(G) = \sum_{\sigma \in \Omega} e^{\beta \sum_{(u,v) \in \mathcal{E}} \mathbf{1}_{\{\sigma(u)=\sigma(v)\}}}.$$

We can similarly define the Swendsen-Wang dynamic with parameter $p \in [0,1]$ on G as the Markov chained defined by first performing p-bond percolation on the graphs spanned by the vertices of each state $\{1, \ldots, q\}$, and then color all the components obtained this way uniformly from the q colors and independently for each cluster. In a similar fashion the Potts model is the stationary measure for this chain.

THEOREM 2.2. *The mixing time of the Swendsen-Wang process for the q-state ferromagnetic Potts model at any temperature on any tree with n vertices is $O(\log n)$, where the constants may depend only on the temperature.*

2.1. Random graph estimates

Due to the percolative nature of the dynamics we require several estimates about percolation on the complete graph. Recall that $G(m, p)$ is the random graph obtained from the complete graph on m vertices by retaining each edge with probability p and erasing it with probability $1 - p$, independently of all other edges. There is rich literature about this model (see [16] and the references within). In particular, there is an interesting phase transition when $p = \frac{1}{m}$. In this work we required several estimates on the size of the largest connected component $\mathcal{C}_1$ for various values of p. In the following we highlight some of these estimates which are interesting for the random graph community.

Fix $\theta > 1$. It is a well known result of Pittel [28] that for $G(m, p)$ with $p = \frac{\theta}{m}$ we have that $\frac{|\mathcal{C}_1| - \beta m}{\sqrt{m}}$ converges in probability to a normal distribution, where β is the unique positive solution of the equation

$$1 - e^{-\theta x} = x.$$

This does not imply, however, the moderate deviation bound on $|\mathcal{C}_1 - \beta m|$

(2.1) $$\mathbf{P}(||\mathcal{C}_1| - \beta m| > A\sqrt{m}) \leq Ce^{-cA^2},$$

for any $A > 0$, which we prove here, see Lemma 5.4.

The study of the random graph "inside" the phase transition was initiated by Bollobás [3] where it is shown that if $p = \frac{1+\epsilon(m)}{m}$ when $\epsilon(m)$ is a positive sequence satisfying $\epsilon(m) \geq m^{-1/3} \log m$, then with high probability $|\mathcal{C}_1| = (2 + o(1))\epsilon m$. The logarithmic corrections were removed by Luczak [22] and this statement holds whenever $\epsilon(m) \gg m^{-1/3}$. A stronger result was recently proved by Pittel and Wormald [29]. They show that in this regime of p, the distribution of $\frac{|\mathcal{C}_1|-2\epsilon m}{\sqrt{m/\epsilon}}$ converges to a normal distribution (this is a corollary of Theorem 6 of [29], but in fact the authors prove much more than this statement).

Surprisingly however, the above results do not give good estimates on $\mathbb{E}|\mathcal{C}_1|$ and on moderate and large deviations of $|\mathcal{C}_1| - 2\epsilon m$. These are crucial in our analysis of the Swendsen-Wang chain since these determine the moments of the increments of the chain. In Section 5 we prove several such estimates, for instance
$$\mathbb{E}|\mathcal{C}_1| = 2\epsilon m + O(\epsilon^{-2} + \epsilon^2 m),$$
see the more accurate inequality in Theorem 5.8. Another interesting estimate is a bound on the deviation of $|\mathcal{C}_1|$,
$$\mathbf{P}\left(\left||\mathcal{C}_1| - 2\epsilon m\right| > A\sqrt{\frac{m}{\epsilon}}\right) \leq Ce^{-cA^2},$$
for any A satisfying $0 \leq A \leq \sqrt{\epsilon^3 m}$. See Theorem 5.9.

CHAPTER 3

Mixing time preliminaries

The following are classical estimates, see [**21**] for further details.

LEMMA 3.1. *Suppose (X_t, Y_t) is a coupling of two copies of the same Markov chain with $X_0 = x$, $Y_0 = y$. We have*

(3.1) $$\|\mathbf{p}^t(x, \cdot) - \mathbf{p}^t(y, \cdot)\|_{TV} \leq \mathbf{P}(X_t \neq Y_t).$$

One can apply triangle inequality to (3.1) to get

$$\|\mathbf{p}^t(x, \cdot) - \pi(\cdot)\|_{TV} \leq \max_y \mathbf{P}(X_t \neq Y_t),$$

and by taking maximum over all x, we have

(3.2) $$\max_x \|\mathbf{p}^t(x, \cdot) - \pi(\cdot)\|_{TV} \leq \max_{x,y} \mathbf{P}(X_t \neq Y_t).$$

This gives the following theorem, which we will use as a main technique to get the mixing time.

LEMMA 3.2. *Assume that for every two states $x, y \in \Omega$ there exists a coupling $\{X_t, Y_t\}$ with $X_0 = x$, $Y_0 = y$ such that $X_t = Y_t$ for some $t \leq L$ with a constant probability $\epsilon > 0$. Then, $T_{\mathrm{mix}}(X, 1-\epsilon) \leq L$.*

As long as one can obtain the order of an upper bound of $T_{\mathrm{mix}}(1-\epsilon)$, the same order of upper bound holds for $T_{\mathrm{mix}}(1/4)$. We refer to Section 4.4 and 4.5 of [**21**] of detailed discussion on this.

CHAPTER 4

Outline of the proof of Theorem 2.1

Due to the length of the proof of Theorem 2.1, we provide here a "road-map" of whole argument. The reader is advised to follow this outline to get the general idea of the proof and go to the main contents for further details whenever needed. Let $\{\sigma_t\}_{t=0}^{\infty}$ be the SW Markov chain. Consider the chain X_t defined by

$$(4.1) \qquad X_t = \left|\sum_v \sigma_t(v)\right|.$$

Since the underlying graph is complete, $\{X_t\}$ is a Markov chain with state space $\{0, \ldots, n\}$. Given X_0, the random variable X_1 is determined by two independently drawn random graphs $G(\frac{n+X_0}{2}, \frac{c}{n})$ and $G(\frac{n-X_0}{2}, \frac{c}{n})$. If we denote by $\{\mathcal{C}_j^+\}_{j\geq 1}$ and $\{\mathcal{C}_j^-\}_{j\geq 1}$ the connected components of the corresponding two random graphs, then X_1 is distributed as

$$(4.2) \qquad \left|\sum_{j\geq 1} \epsilon_j |\mathcal{C}_j^+| + \sum_{j\geq 1} \epsilon_j' |\mathcal{C}_j^-|\right|,$$

where $\{\epsilon_j\}$ and $\{\epsilon_j'\}$ are i.i.d. random variables taking 1 with probability $1/2$ and -1 otherwise. This is the reason that the moments and large deviation estimates of random graph component sizes are useful in our approach.

Frequently, to obtain upper bounds on the mixing time of the SW chain we will obtain a bound on the mixing time of the chain X_t and then use the following lemma, which appears in [7] and is based on the path coupling idea of Bubley and Dyer [6].

LEMMA 4.1. *Suppose $\{\sigma_t\}$ and $\{\sigma_t'\}$ are two SW chains such that $X_0 = X_0'$. There exists a coupling of the two chains such that with probability at least $\frac{1}{2}$ the two chains meet after $O(\log n)$ steps.*

4.1. Outline of the Proof of Theorem 2.1 (i)

By Lemmas 3.2 and 4.1 it suffices to show that we can couple two copies of the magnetization chain X_t and X_t' such that they meet in $O(\log n)$ with probability $\Omega(1)$ which is uniform over all initial values x_0 and x_0'. It turns out that the stationary distribution is concentrated in a window of length $\sqrt{n}$ around $\gamma_0 n$ for some $\gamma_0 = \gamma_0(c) \in [0, 1]$. In fact, the one step evolution of X_t essentially contracts the second moment of $|X_t - \gamma_0 n|$. That is, we have

$$(4.3) \qquad \mathbb{E}(X_1 - \gamma_0 n)^2 \leq \delta(x_0 n - \gamma_0 n)^2 + Bn,$$

for some constants $\delta \in (0, 1)$ and large B. See Theorem 6.2. It follows quickly that there exists an interval of values $I = [\gamma_0 n - A\sqrt{n}, \gamma_0 n + A\sqrt{n}]$ for some large constant A such that for any initial value x_0 we have that $X_t \in I$ with probability $\Omega(1)$ whenever $t = \Theta(\log n)$.

Once the two chains are both in the interval I one can show that they can be coupled to meet in the next step with probability $\Omega(1)$. This is the content of Theorem 6.5. The main idea of that argument is that the random graph $G(n, \frac{c}{n})$ has $\Theta(n)$ isolated vertices with high probability and that the difference of the sums of the spins of the two chains *before* we assign spins to the isolated vertices is $O(\sqrt{n})$. Thus, we one can couple the two chains to correct the $O(\sqrt{n})$ error by assignment of those isolated vertices. This follows from the classical local central limit theorem for the simple random walk.

4.2. Outline of the Proof of Theorem 2.1 (iii)

Since we need to prove an $O(1)$ upper bound we cannot use Lemma 4.1 here. However, the study of X_t's evolution will still be useful. As in the supercritical case, the stationary measure is concentrated in a window of width $\Theta(\sqrt{n})$, but this time around 0 and mixing occurs much faster. We will show, as before, that we have contraction, that is,

$$(4.4) \qquad \mathbb{E}(X_1^2 \mid X_0) \leq \delta X_0^2 + Bn$$

for some constants $\delta \in (0,1)$ and large B and for all $x_0 \in [0,1]$. Moreover, if $0 \leq X_0 \leq \frac{1}{c} - \frac{1}{2}$, we have

$$(4.5) \qquad \mathbb{E} X_1^2 \leq Bn,$$

see Theorem 7.2. The first inequality implies that X_t will be in the window $[0, (\frac{1}{c} - \frac{1}{2})n]$ in $O(1)$ steps with probability $\Omega(1)$. The second inequality implies that from this window X_t jumps into $[0, A\sqrt{n}]$ with high probability in just one more step. This gives the mixing time upper bound on the chain X_t.

To go further and obtain a mixing time of the SW chain one needs to consider the following two-dimensional chain. For a starting configuration σ_0, let G_1 denote the vertices with positive spin and G_2 be its complement. Let (Y_t, Z_t) be a two-dimensional Markov chain, where Y_t records the number of vertices with positive spin in G_1 and Z_t records the number vertices with positive spin in G_2. By symmetry, the probability of the SW chain of being at σ at time t is the same for all σ which have the same two-dimensional chain value. Consequently, the total variation distance of σ_t from stationarity is the same as the total variation distance of the two-dimensional chain from its stationary distribution. By Lemma 3.2, it suffices to provide a coupling of such two-dimensional chains so that they meet in $O(1)$ steps with probability $\Omega(1)$. By our previous argument, $Y_t + Z_t$ will be in the window $[\frac{n}{2} - A\sqrt{n}, \frac{n}{2} + A\sqrt{n}]$ within $O(1)$ steps. One can show that once inside this window, such a coupling does exist. See Proposition 7.3. The idea is similar to the proof of part (i) by considering the isolated vertices in the two random graphs.

4.3. Outline of the Proof of Theorem 2.1 (ii)

We again use Lemma 4.1 and Lemma 3.2 to bound the mixing time of the magnetization chain. However, to simplify our calculations, we will consider a slight modification to the magnetization chain X_t. Instead of choosing a random spin for each component after the percolation step, we assign a positive spin to the *largest* component and random spins for all other components. Let X'_t be the sum

of spins at time t (notice that we do *not* take absolute values here), that is,
(4.6)
$$X'_{t+1} \overset{d}{=} \max\{|\mathcal{C}_1^+(t)|, |\mathcal{C}_1^-(t)|\} + \epsilon \min\{|\mathcal{C}_1^+(t)|, |\mathcal{C}_1^-(t)|\} + \sum_{j \geq 2} \epsilon_j |\mathcal{C}_j^+(t)| + \sum_{j \geq 2} \epsilon'_j |\mathcal{C}_j^-(t)|,$$

where as usual $\epsilon, \{\epsilon_j\}$ and $\{\epsilon'_j\}$ are independent mean zero $\pm$ signs. This chain has state space $\{-n, \ldots, n\}$ and its absolute value has the same distribution as our original chain X_t. As a consequence, any upper bound on the mixing time of X'_t implies the same upper bound on the mixing time of X_t. For convenience, we will now denote this modified chain by X_t. Let X_t and Y_t be two such chains such that X_t start from an arbitrary location X_0 and Y_t starts from the stationary distribution. We will show that we can couple X_t and Y_t so that they meet in $O(n^{1/4})$ steps with probability $\Omega(1)$. It will become evident that it suffices to restrict the attention to $X_0 \in [0, n]$. We will divide this into two subcases:

(i) $X_0 \in [n^{3/4}, n]$,
(ii) $X_0 \in [0, n^{3/4}]$,

and consider them separately, let us begin with case (i). In this case the coupling strategy is as follows.

Consider the first *crossing time* of X_t and Y_t, that is, the first time t such that $\text{sign}(X_t - Y_t) \neq \text{sign}(X_{t-1} - Y_{t-1})$. We will show that this is likely to occur only when the two chains take values $\Theta(n^{3/4})$ and, more importantly, the distance between the chains one step before the crossing time is of order $n^{5/8}$. This is the content of Theorem 8.5. The fact that one time step before the crossing time is not a stopping time is problematic and requires an *overshoot* estimate stating that the two chains are not likely to cross each other from distance larger than $O(n^{5/8})$. For random walks, these kind of estimates are classical (see for instance [19]). The key estimate here is Theorem 8.8.

Next we show that when the chains take values $\Theta(n^{3/4})$ they satisfying a local central limit theorem in scale $n^{5/8}$. In particular, the chain has probability $\Omega(n^{-5/8})$ to move to any point x in an interval of size $\Theta(n^{5/8})$ around the starting point. We use the standard characteristic function technique to show this, see Lemma 8.19. Now we are ready to conclude the proof in this case since we know that a step before the crossing time the chains have already been at distance $O(n^{5/8})$ from each other, so the local CLT provides a way to couple them in a few additional steps after the crossing time. See the proof of Theorem 8.1.

Let us consider now case (ii) in which $X_0 \in [0, n^{3/4}]$. To handle this case define $I = [-An^{2/3}, An^{2/3}]$ and proceed in two steps.

(1) With high probability X_t will visit the interval I by time $O(n^{1/4})$. This is proved in Theorem 8.24 and is based on the fact that the drift of the chain $|X_t|$ in this regime is approximately $-n^{1/2}$ (this is a small negative drift).
(2) Once the chain is inside I, it will be pushed above $\Omega(n^{3/4})$ within $O(n^{1/4})$ steps. See Theorem 8.23.

Thus, with these two claims we see that in at most $O(n^{1/4})$ steps the chain is pushed into the $n^{3/4}$ regime and we may use the theorems of case (i) to conclude. Let us briefly expand on the proofs of (1) and (2).

The proof of (1) relies on the fact that the chain has a negative drift of magnitude $\Omega(n^{1/2})$ as long as $X_t \notin I$. This follows rather easily from the random graph estimate Theorem 5.15. Note, however, that Theorem 5.15 estimates the expected size of the cluster discovered in time $\delta \epsilon m$ in the exploration process for some small $\delta > 0$ and *not* of the largest cluster $\mathcal{C}_1$. We denote the former cluster by $\mathcal{C}_{\delta \epsilon m}$ and remark that it has high probability of being the largest. However, we were unable to prove the estimate of Theorem 5.15 for $\mathcal{C}_1$ but only for $\mathcal{C}_{\delta \epsilon m}$. This is the reason we need to consider yet another slight modification of the magnetization chain: instead of giving a plus sign to $\mathcal{C}_1$ and drawing random signs for the rest of the clusters, we give the plus sign to $\mathcal{C}_{\delta \epsilon m}$ and the rest receive random signs. From this point on the proof of (1) is rather straightforward.

For the proof of (2) one has to show that when X_t is in I, even though the drift is negative there is still enough noise to eventually push X_t to the $n^{3/4}$ regime. We were unable to pursue this strategy since it involves very delicate random graph estimates we were unable to obtain. Instead we use the following coupling idea. Since the stationary distribution normalized by $n^{3/4}$ has a weak limit with positive density at 0, the expected number of visits to I by the stationary chain before time T is $\Theta\bigl(\frac{n^{2/3}}{n^{3/4}} T\bigr)$. In Lemma 8.25 we show that when $T = \Theta(n^{1/4})$ the actual number of visits to I is positive with high probability. Next, to show that X_t is pushed upwards we start a stationary chain Z_t and wait until it enters I. We then couple X_t and Z_t such that they meet inside I and from that point they stay together. The only technical issue with this strategy is how to perform the coupling of Z_t and X_t inside I. This will follow, as before, from a uniform lower bound stating that for any $x, x_0 \in I$ we have

$$\mathbf{P}(X_1 = x \mid X_0 = x_0) \geq c n^{-2/3}.$$

This estimate is done inside the scaling window of the random graph phase transition and so the proofs are different from the previous ones and require some combinatorial estimates. See Lemmas 8.26 and Lemma 8.27.

CHAPTER 5

Random graph estimates

In this section we prove some facts about random graphs which will be used in the proof. These lemmas might also be of sperate interests in random graph theory. Recall that $G(m,p)$ is obtained from the complete graph on m vertices by retaining independently each edge with probability p and deleting it with probability $1-p$. We denote by $\mathcal{C}_j$ the j-th largest component of $G(m,p)$.

5.1. The exploration process

We recall an exploration process, due to Karp and Martin-Löf (see [18] and [23]), in which vertices will be either *active, explored* or *neutral*. After the completion of step $t \in \{0, 1, \ldots, m\}$ we will have precisely t explored vertices and the number of the active and neutral vertices is denoted by A_t and N_t respectively. Fix an ordering of the vertices $\{v_1, \ldots, v_m\}$. In step $t = 0$ of the process, we declare vertex v_1 active and all other vertices neutral. Thus $A_0 = 1$ and $N_0 = m - 1$. In step $t \in \{1, \ldots, m\}$, if $A_{t-1} > 0$, then let w_t be the first active vertex; if $A_{t-1} = 0$, let w_t be the first neutral vertex. Denote by η_t the number of neutral neighbors of w_t in $G(m,p)$, and change the status of these vertices to active. Then, set w_t itself explored.

Denote by $\mathcal{F}_t$ the σ algebra generated by $\{\eta_1, \ldots, \eta_t\}$. Observe that given $\mathcal{F}_{t-1}$ the random variable η_t is distributed as $\text{Bin}(N_{t-1} - \mathbf{1}_{\{A_{t-1}=0\}}, p)$ and we have the recursions

(5.1) $$N_t = N_{t-1} - \eta_t - \mathbf{1}_{\{A_{t-1}=0\}}, \qquad t \leq m,$$

and

(5.2) $$A_t = \begin{cases} A_{t-1} + \eta_t - 1, & A_{t-1} > 0 \\ \eta_t, & A_{t-1} = 0, \end{cases} \qquad t \leq m.$$

As every vertex is either neutral, active or explored,

(5.3) $$N_t = m - t - A_t, \qquad t \leq m.$$

At each time $j \leq m$ in which $A_j = 0$, we have finished exploring a connected component. Hence the random variable Z_t defined by

$$Z_t = \sum_{j=1}^{t-1} \mathbf{1}_{\{A_j=0\}},$$

counts the number of components completely explored by the process before time t. Define the process $\{Y_t\}$ by $Y_0 = 1$ and

$$Y_t = Y_{t-1} + \eta_t - 1.$$

By (5.2) we have that $Y_t = A_t - Z_t$, i.e. Y_t counts the number of active vertices at step t minus the number of components completely explored before step t.

LEMMA 5.1. *For any t we have*

$$Y_t \overset{d}{\leq} \text{Bin}(m-1, 1-(1-p)^t) + 1 - t, \tag{5.4}$$

and

$$Y_t \overset{d}{\geq} \text{Bin}(m-t-1, 1-(1-p)^t) + 1 - t. \tag{5.5}$$

Proof. For each vertex v at each step of the process we examine precisely one of its edges emanating from it unless the vertex is active or explored at this step. Thus, all the vertices for which the process discovered an open edge emanating from them between time 1 and t are active, except for at least t of them which are explored. The probability of a vertex having no open edges explored from it between time 1 and t is precisely $(1-p)^t$. This shows (5.4).

The reason this bound is not precise is that it is possible that a neutral vertex turns to be active because there were no more active vertices at this step of the exploration process. This, however, can only happen at most t times between time 1 and t and this gives the lower bound (5.5). □

At each step we marked as explored precisely one vertex. Hence, the component of v_1 has size $\min\{t \geq 1 : A_t = 0\}$. Moreover, let $t_1 < t_2 \ldots$ be the times at which $A_{t_j} = 0$; then $(t_1, t_2 - t_1, t_3 - t_2, \ldots)$ are the sizes of the components. Observe that $Z_t = Z_{t_j} + 1$ for all $t \in \{t_j + 1, \ldots, t_{j+1}\}$. Thus $Y_{t_{j+1}} = Y_{t_j} - 1$ and if $t \in \{t_j+1, \ldots, t_{j+1}-1\}$ then $A_t > 0$, and thus $Y_{t_{j+1}} < Y_t$. By induction we conclude that $A_t = 0$ if and only if $Y_t < Y_s$ for all $s < t$. In other words $A_t = 0$ if and only if $\{Y_t\}$ has hit a new record minimum at time t. By induction we also observe that $Y_{t_j} = -(j-1)$ and that for $t \in \{t_j+1, \ldots t_{j+1}\}$ we have $Z_t = j$. Also, by our previous discussion for $t \in \{t_j+1, \ldots t_{j+1}\}$ we have $\min_{s \leq t-1} Y_s = Y_{t_j} = -(j-1)$, hence by induction we deduce that $Z_t = -\min_{s \leq t-1} Y_s + 1$. Consequently,

$$A_t = Y_t - \min_{s \leq t-1} Y_s + 1. \tag{5.6}$$

LEMMA 5.2. *For all $p \leq \frac{2}{m}$ there exists a constant $c > 0$ such that for any integer $t > 0$,*

$$\mathbf{P}\left(N_t \leq m - 5t\right) \leq e^{-ct}.$$

Where we recall N_t is the number of neutral points in exploration process at time t.

The proof of Lemma 5.2 can be found in Lemma 3 of [**24**].

5.2. Random graph lemmas for non-critical cases

Let $G(m, p)$ be the random graph where $p = \frac{\theta}{m}$.

LEMMA 5.3. *Suppose $\theta < 1$ is a constant. Then we have*

$$\mathbb{E}(\sum_{j \geq 1} |\mathcal{C}_j|^2) \leq \frac{m}{1-\theta}. \tag{5.7}$$

Proof. Observe that $\sum_{j \geq 1} |\mathcal{C}_j|^2 = \sum_v |\mathcal{C}(v)|$ since in the right hand side each component $\mathcal{C}(v)$ is counted precisely $|\mathcal{C}(v)|$ times. Hence

$$\mathbb{E}(\sum_{j \geq 1} |\mathcal{C}_j|^2) = \mathbb{E} \sum_v |\mathcal{C}(v)| = m E |\mathcal{C}(v)|. \tag{5.8}$$

In the exploration process, we can couple Y_t with a process W_t with i.i.d. increment of $\text{Bin}(m, \theta/m) - 1$ and $W_0 = 1$ such that $W_t \geq Y_t$. Thus, the hitting time of 0 for Y_t which equals to $|\mathcal{C}(v)|$ is bounded from above by the hitting time of 0 for W_t. For W_t, we have $\mathbb{E}\tau = 1/(1-\theta)$ by Wald's Lemma. This concludes the proof. □

For $\theta > 1$ let $\beta = \beta(\theta)$ be the unique positive solution of the equation

(5.9) $$1 - e^{-\theta x} = x.$$

In [28] it was proved that $\frac{|\mathcal{C}_1| - \beta m}{\sqrt{m}}$ converges in distribution to a normal distribution. We were unable to deduce from that result moderate deviation estimates, and we provide them in the following lemma.

LEMMA 5.4. *There exists constants $c = c(\theta) > 0$ and universal constant C such that for any $A > 0$ we have*

(5.10) $$\mathbf{P}(||\mathcal{C}_1| - \beta m| \geq A\sqrt{m}) \leq Ce^{-cA^2}.$$

Proof. Assume $A \leq \sqrt{m}$ otherwise this probability is 0. Let $\xi = \xi(\theta) > 0$ be a large constant that we will determine later. We will show that for some $c > 0$

(5.11) $$\mathbf{P}(Y_{\beta m + A\sqrt{m}} \geq -cA\sqrt{m}) \leq e^{-cA^2},$$

and that

(5.12) $$\mathbf{P}\Big(\bigcup_{cA\sqrt{m} \leq t \leq \beta m - \xi A\sqrt{m}} Y_t < 0\Big) \leq Ce^{-cA^2}.$$

If these two events do not occur, then there exists a component of size in $[\beta m - (\xi + c)A\sqrt{m}, \beta m + A\sqrt{m}]$. The remaining graph is a subcritical random graph and it is a classical result that the probability that it contains a component of size $\Theta(m)$ decays exponentially in m, and this will conclude the proof.

The proof of (5.11) is based on the stochastic upper bound of Y_t in (5.4). Plugging in $t = \beta m + A\sqrt{m}$ and using the fact that $1 - x \geq e^{-x-x^2}$ for small enough x we get

$$\begin{aligned}\mathbf{P}(Y_{\beta m + A\sqrt{m}} \geq -cA\sqrt{m}) &\leq \mathbf{P}\Big(\text{Bin}(m, 1-(1-\frac{\theta}{m})^{\beta m + A\sqrt{m}}) \geq \beta m + (1-c)A\sqrt{m}\Big) \\ &\leq \mathbf{P}\Big(\text{Bin}(m, 1 - e^{-\theta\beta - A\theta m^{-1/2} - A\theta^2 m^{-3/2}}) \geq \beta m + (1-c)A\sqrt{m}\Big).\end{aligned}$$

A quick calculation using the fact that $1 - e^{-\theta\beta} = \beta$ gives that the expected value of this binomial random variable is at most

$$\beta m + A\theta e^{-\theta\beta}\sqrt{m} + O(1).$$

Since $\theta e^{-\theta\beta} < 1$ it follows that we can choose c so small so that this expectation is less than $\beta m + (1-2c)A\sqrt{m}$, and then Azuma-Hoeffding inequality (see for instance Theorem 7.2.1 of [1]) gives that

$$\mathbf{P}(Y_{\beta m + cA\sqrt{m}} \geq -A\sqrt{m}) \leq e^{-cA^2}.$$

We now turn to prove (5.12). For this we will divide $[cA\sqrt{m}, \beta m - \xi A\sqrt{m}]$ into two subintervals $[cA\sqrt{m}, \delta\beta m]$ and $[\delta\beta m, \beta m - \xi A\sqrt{m}]$ where $\delta > 0$ is a small constant that will be chosen later. For convenience write $\alpha = \frac{t}{m}$. For any $t \in$

$[cA\sqrt{m}, \delta\beta m]$ we have by (5.5) and the fact that $1 - x \le e^{-x}$ for all $x \ge 0$ that

$$\begin{aligned}\mathbf{P}(Y_t < 0) &\le \mathbf{P}\Big(\mathrm{Bin}((1-\alpha)m, 1 - (1-\frac{\theta}{m})^{\alpha m}) \le \alpha m\Big) \\ &\le \mathbf{P}\Big(\mathrm{Bin}((1-\alpha)m, 1 - e^{-\theta\alpha})) \le \alpha m\Big).\end{aligned}$$

Since $1 - e^{-x} \ge x - x^2$ for all $x \ge 0$ we deduce that the expectation of the last binomial is at least

$$(1-\alpha)(\theta\alpha - \theta^2\alpha^2) > \alpha,$$

since $\theta > 1$ when $\alpha = t/m \le \delta$ and $\delta = \delta(\theta) > 0$ is chosen small enough. By a standard large deviation estimate (see for instance, Corollary A.1.14 of [1]), we have that

$$\mathbf{P}(Y_t < 0) \le e^{-c\alpha m},$$

for some $c = c(\theta)$ and all $t \in [cA\sqrt{m}, \delta\beta m]$. It follows from the union bound that

(5.13) $$\mathbf{P}(\bigcup_{cA\sqrt{m} \le t \le \delta\beta m} Y_t < 0) = O(e^{-cA\sqrt{m}}).$$

For the interval $[\delta\beta m, \beta m - \xi A\sqrt{m}]$ we will use the process $\widetilde{Y}_t$ which approximates Y_t introduced by Bollobas and Riordan [4]. We write

$$D_t = \mathbb{E}(\eta_t - 1 | F_{t-1}),$$

and define

$$\Delta_t = \eta_t - 1 - D_t.$$

Let $y_t = m - t - m(1-p)^t$ and define the approximation process by

(5.14) $$\widetilde{Y}_t = y_t + \sum_{i=1}^{t}(1-p)^{t-i}\Delta_i.$$

In [4] Lemma 3 it is proved that for any $p > 0$ and any $1 \le t \le m$ we have

(5.15) $$|Y_t - \widetilde{Y}_t| \le pt Z_t.$$

Put $\tau = \min\{t \ge \delta\beta m, A_t = 0\}$. We have

(5.16) $$\begin{aligned}\mathbf{P}(\tau < \beta m - \xi A\sqrt{m}) &\le \mathbf{P}(|Y_\tau - \widetilde{Y}_\tau| \ge \theta A\sqrt{m}) \\ &+ \mathbf{P}(|Y_\tau - \widetilde{Y}_\tau| < \theta A\sqrt{m}, \tau < \beta m - \xi A\sqrt{m}).\end{aligned}$$

By (5.15) the first term has the upper bound

$$\mathbf{P}(|Y_\tau - \widetilde{Y}_\tau| \ge \theta A\sqrt{m}) \le \mathbf{P}(Z_\tau \ge A\sqrt{m}) = O(e^{-cA\sqrt{m}}),$$

since $Z_\tau \ge A\sqrt{m}$ implies that there exists at least one time t in $[A\sqrt{m}, \delta\beta m]$ such that $Y_t < 0$. The bound follows immediately from our estimate in (5.13).

To bound the second term of (5.16) observe that on $[\delta\beta m, \beta m - \xi A\sqrt{m}]$, the minimum of y_t is attained at the right end of the interval with value $(1 - \theta e^{-\theta\beta})\xi A\sqrt{m}(1 + o(1))$. Thus if we choose $\xi = \xi(\theta)$ large enough such that $(1 - \theta e^{-\theta\beta})\xi > \theta$ and write $c = (1 - \theta e^{-\theta\beta})\xi - \theta$, we have

(5.17) $\mathbf{P}(|Y_\tau - \widetilde{Y}_\tau| < \theta A\sqrt{m}, \tau < \beta m - \xi A\sqrt{m}) \le \mathbf{P}\big(\sum_{i=1}^{\tau}(1-p)^{\tau-i}\Delta_i < -cA\sqrt{m}\big),$

since $Y_\tau \leq 0$ by definition. Notice that τ is at most m, thus it suffices to bound from above
$$\mathbf{P}\Big(\max_{1\leq t\leq m}(-\sum_{i=1}^{t}(1-p)^{t-i}\Delta_i) > cA\sqrt{m}\Big).$$
Notice that $(1-p)^t = \Theta(1)$, hence it is equivalent to bound
$$\mathbf{P}\Big(\max_{1\leq t\leq m}(-\sum_{i=1}^{t}(1-p)^{-i}\Delta_i) > cA\sqrt{m}\Big).$$
Let $a > 0$ be a small number (eventually we will take $a = \Theta(m^{-1/2})$). Direct computation and the fact that $1 + x > e^{x-x^2}$ for negative x when $|x|$ is small enough and some Taylor expansion yield
$$\mathbb{E}(e^{-a(1-p)^{-i}\Delta_i}|F_{i-1}) = (1+p(e^{-a(1-p)^{-i}}-1))^{N_{i-1}-\mathbf{1}_{\{A_{i-1}=0\}}}e^{a(1-p)^{-i}p(N_{i-1}-\mathbf{1}_{\{A_{i-1}=0\}})}$$
$$(5.18) \qquad \geq e^{\frac{a^2 p}{3}(N_{i-1}-\mathbf{1}_{\{A_{i-1}=0\}})(1-p)^{-2i}} \geq 1,$$
when a is small enough. Thus we conclude $e^{-a\sum_{i=1}^{t}(1-p)^i\Delta_i}$ is a submartingale. By Doob's maximal inequality (see [**10**]) we have
$$\mathbb{E}\Big(\max_{1\leq t\leq m}e^{-a\sum_{i=1}^{t}(1-p)^i\Delta_i}\Big)^2 \leq 4\mathbb{E}e^{-2a\sum_{i=1}^{m}(1-p)^i\Delta_i}.$$
On the other hand, the fact that $1 + x \leq e^x$ for all x yields
$$(5.19) \quad \mathbb{E}(e^{-a(1-p)^{-i}\Delta_i}|F_{i-1}) \leq e^{(\frac{a^2}{2}+O(a^3))p(N_{i-1}-\mathbf{1}_{\{A_{i-1}=0\}})(1-p)^{-2i}}.$$
Since $N_t \leq m$ we get
$$\mathbb{E}e^{-2a\sum_{i=1}^{m}(1-p)^i\Delta_i} \leq 4e^{Cma^2},$$
where $C = C(\theta)$. By Markov's inequality, we have
$$\mathbf{P}\Big(\max_{1\leq t\leq m}(-\sum_{i=1}^{t}(1-p)^{-i}\Delta_i) > cA\sqrt{m}\Big) \leq e^{Cma^2 - 2acA\sqrt{m}}.$$
Choosing $a = \frac{cA\sqrt{m}}{Cm}$ to minimize the right hand side, we conclude
$$\mathbf{P}\Big(\max_{1\leq t\leq m}(-\sum_{i=1}^{t}(1-p)^{-i}\Delta_i) > \theta A\sqrt{m}\Big) \leq 4e^{-cA^2},$$
for some $c = c(\theta)$ which is a continuous function of θ, and this concludes the proof. $\square$

COROLLARY 5.5. *Suppose $\theta \in [a,b]$ where $a > 1$. Then, there exists a constant $C = C(a,b)$ such that for $G(m, \frac{\theta}{m})$ we have*
$$(5.20) \qquad \Big|\mathbb{E}|\mathcal{C}_1| - \beta(\theta)m\Big| \leq C\sqrt{m},$$

Proof. It follows immediately by integrating Lemma 5.4. $\square$

COROLLARY 5.6. *Suppose $\theta \in [a,b]$ where $a > 1$. There exists a constant $C = C(a,b)$ such that for $G(m, \frac{\theta}{m})$ we have*
$$(5.21) \qquad \mathbb{E}(\sum_{j\geq 1}|\mathcal{C}_j|^2) \leq (\mathbb{E}|\mathcal{C}_1|)^2 + Cm.$$

Proof. Notice that

(5.22) $$\mathbb{E}\sum_{j\geq 1}|\mathcal{C}_j|^2 - (\mathbb{E}|\mathcal{C}_1|)^2 = \left(\mathbb{E}|\mathcal{C}_1|^2 - (\mathbb{E}|\mathcal{C}_1|)^2\right) + \mathbb{E}\sum_{j\geq 2}|\mathcal{C}_j|^2.$$

We have that

$$\mathbb{E}|\mathcal{C}_1|^2 - (\mathbb{E}|\mathcal{C}_1|)^2 = \mathbb{E}(|\mathcal{C}_1| - \mathbb{E}|\mathcal{C}_1|)^2 \leq \mathbb{E}(|\mathcal{C}_1| - \beta m)^2.$$

By integrating Lemma 5.4 we get

$$\mathbb{E}|\mathcal{C}_1|^2 - (\mathbb{E}|\mathcal{C}_1|)^2 \leq Cm.$$

For supercritical random graph $G(m, \frac{\theta}{m})$, it is a classical result that $|\mathcal{C}_1| \in ((\beta - \epsilon)m, (\beta + \epsilon)m)$ with probability at least $1 - e^{-c\epsilon m}$ for fixed ϵ. Conditioned on this and the vertex set of $\mathcal{C}_1$, the other components are distributed as $G(m - |\mathcal{C}_1|, \frac{\theta}{m})$ (which is subcritical) restricted to the event that it does not contain any component larger than $|\mathcal{C}_1|$. This event happens with probability at most e^{-cm}. Thus we obtain

$$\mathbb{E}(\sum_{j\geq 2}|\mathcal{C}_j|^2) \leq (1 + o(1))\frac{m}{1 - (1 - \beta)\theta},$$

□

LEMMA 5.7. *Let $M = \sum_{v \in V}\mathbf{1}_{\{v \text{ is isolated}\}}$ be the number of isolated vertices in $G(m, \theta/m)$ where $\theta > 0$ is a constant. There exists a constant $C > 0$ such that*

$$\mathbf{P}(M \geq Cm) = 1 - O(\frac{1}{m}).$$

Proof. We have

$$\mathbb{E}M = \sum_{v\in V}\mathbf{P}(v \text{ is isolated.}) = m(1 - \frac{\theta}{m})^{m-1},$$

and

$$\mathbb{E}M^2 = \sum_{v\in V}\mathbf{P}(v \text{ is isolated.}) + \sum_{v,w\in V}\mathbf{P}(v, w \text{ are both isolated.})$$
$$= m(1 - \frac{\theta}{m})^{m-1} + m(m-1)(1 - \frac{\theta}{m})^{2m-3}.$$

Thus, we obtain

$$\mathbb{E}(M - \mathbb{E}M)^2 = O(m).$$

By Markov's inequality,

$$\mathbf{P}(M \leq \frac{1}{2}\mathbb{E}M) \leq \mathbf{P}\left((M - \mathbb{E}M)^2 \geq \frac{1}{4}(\mathbb{E}M)^2\right) \leq \frac{\mathbb{E}(M - \mathbb{E}M)^2}{\frac{1}{4}(\mathbb{E}M)^2} = O(\frac{1}{m}).$$

Since $\mathbb{E}M = \Theta(m)$, we finished the proof. □

5.3. Random graph lemmas for the near-critical case

In [**29**], Pittel and Wormald study the near-critical random graph $G(m,p)$ where $p = \frac{1+\epsilon}{m}$ with $\epsilon = o(1)$ but $\epsilon^3 m \to \infty$. A direct corollary of Theorem 6 of [**29**] shows that in this regime $\frac{|\mathcal{C}_1|-2\epsilon m}{\sqrt{m/\epsilon}}$ converges in distribution to a normal random variable (see also [**4**] for a recent simple proof of this fact), and that a local central limit theorem holds. Unfortunately, one cannot deduce from that precise bounds on the average size of $|\mathcal{C}_1|$ and moderate deviations estimates on $|\mathcal{C}_1| - 2\epsilon m$. The following two theorems give these estimates.

THEOREM 5.8. *Consider $G(m,p)$ with $p = \frac{1+\epsilon}{m}$ where $\epsilon = o(1)$ and there exists a large constant $A > 0$ such that $\epsilon^3 m \geq A \log m$. Then we have that*

$$\mathbb{E}|\mathcal{C}_1| \leq 2\epsilon m - \frac{8}{3}\epsilon^2 m + O(\epsilon^3 m),$$

and there exists a constant $C > 0$ such that

$$\mathbb{E}|\mathcal{C}_1| \geq 2\epsilon m - C(\epsilon^{-2} + \epsilon^2 m).$$

THEOREM 5.9. *Consider $G(m,p)$ with $p = \frac{1+\epsilon}{m}$ where $\epsilon^3 m \geq 1$. Then there exists some $c > 0$ such that*

$$\mathbf{P}\left(\big||\mathcal{C}_1| - 2\epsilon m\big| > A\sqrt{\frac{m}{\epsilon}}\right) = O\left(e^{-cA^2}\right),$$

for any A satisfying $2 \leq A \leq \sqrt{\epsilon^3 m}/10$.

COROLLARY 5.10. *Consider $G(m,p)$ with $p = \frac{1+\epsilon}{m}$ where $\epsilon^3 m \geq 1$, then*

$$\mathbb{E}\big||\mathcal{C}_1| - 2\epsilon m\big|^k \leq C\left(\frac{m}{\epsilon}\right)^{k/2}.$$

THEOREM 5.11. *For any large constant A and small $\delta > 0$ there exists a constant $q_1(A, \delta) > 0$ such that the following hold. Consider $G(m,p)$ with $p = \frac{1+\epsilon}{m}$ where $\epsilon \in [A^{-1} m^{-1/4}, A m^{-1/4}]$, then*

$$\mathbf{P}\big(|\mathcal{C}_1| \in [2\epsilon m - \delta m^{5/8}, 2\epsilon m + \delta m^{5/8}]\big) \geq q_1 > 0.$$

Theorem 5.11 is a direct corollary of Theorem 6 of [**29**] which provides a central limit theorem for the giant component. Next we provide some moment estimates of component sizes in the subcritical and supercritical regime.

THEOREM 5.12. *Consider $G(m,p)$ with $p = \frac{1-\epsilon}{m}$ and $\epsilon^3 m \geq 1$. Then we have*

(i) $\mathbb{E} \sum_{j \geq 1} |\mathcal{C}_j|^k = O(m\epsilon^{-2k+3})$ *for any fixed $k \geq 2$,*
(ii) $\mathbb{E} \sum_{i,j} |\mathcal{C}_i|^2 |\mathcal{C}_j|^2 = O(m^2 \epsilon^{-2})$,
(iii) $\mathbb{E} \sum_{j \geq 1} |\mathcal{C}_j|^2 \geq c m \epsilon^{-1}$.

THEOREM 5.13. *Consider $G(m,p)$ with $p = \frac{1+\epsilon}{m}$ with $\epsilon^3 m \geq 1$. Then we have*

(i) $\mathbb{E}|\mathcal{C}(v)|^k = O(\epsilon^{k+1} m^k)$, *for any fixed $k \geq 2$.*
(ii) $\mathbb{E} \sum_{j \geq 2} |\mathcal{C}_j|^k = O(m\epsilon^{-2k+3})$,
(iii) $\mathbb{E} \sum_{i,j \geq 2} |\mathcal{C}_i|^2 |\mathcal{C}_j|^2 = O(m^2 \epsilon^{-2})$.

THEOREM 5.14. *Consider $G(m,p)$ with $p = \frac{1-\epsilon}{m}$ where $\epsilon \in [A^{-1}m^{-1/4}, Am^{-1/4}]$. Then, for any small positive constant δ, there exist $K = K(A, \delta)$ and $q_2 = q_2(A, \delta)$ such that*

$$\mathbf{P}\Big(\sum_{|\mathcal{C}_j| \leq \delta\sqrt{m}} |\mathcal{C}_j|^2 \geq Km^{5/4} \Big) \geq q_2 > 0.$$

In the following theorem we derive estimates on the expected cluster size valid as long as $\epsilon^3 m \geq 1$. We believe these estimates should hold for $\mathcal{C}_1$ but we were not able to prove that. The difficulty rises because when $\epsilon^3 m$ is large but does not grow at least logarithmically, it is hard to rule out the possibility that $\mathcal{C}_1$ is discovered after time $\delta\epsilon m$ for some fixed $\delta > 0$. Luckily, for the main proof it suffices to have these estimate for $\mathcal{C}_{\delta\epsilon m}$, that component discovered at time $\delta\epsilon m$, rather than $\mathcal{C}_1$. This becomes evident in the proof of Theorem 8.24.

THEOREM 5.15. *Consider $G(m,p)$ with $p = \frac{1+\epsilon}{m}$ and assume $\epsilon^3 m \geq 1$. For some fixed $\delta > 0$ let $\mathcal{C}_{\delta\epsilon m}$ be the component which is discovered by the exploration process at time $\delta\epsilon m$ (in other words, the length of the excursion of Y_t containing the time $\delta\epsilon m$). Then there is some small value of $\delta > 0$ such that*

(i) $\mathbb{E}|\mathcal{C}_{\delta\epsilon m}| \leq 2\epsilon m - c\epsilon^{-2}$.
(ii) $\mathbb{E}\sum_{\mathcal{C}_j \neq \mathcal{C}_{\delta\epsilon m}} |\mathcal{C}_j|^k \leq Cm\epsilon^{-2k+3}$, for $k = 2, 4$,

where C and c are positive universal constants.

Before proceeding to the proofs of the theorems stated in this section, we first require some preparations about processeses with i.i.d. increments.

5.3.1. Processes with i.i.d. increments. Fix some small $\epsilon > 0$ and let $p = \frac{1+\epsilon}{m}$ for some integer $m > 1$. Let $\{\beta_j\}$ be a sequence of random variables distributed as $\text{Bin}(m, p)$. Let $\{W_t\}_{t \geq 0}$ be a process defined by

$$W_0 = 1, \qquad W_t = W_{t-1} + \beta_t - 1.$$

Let τ be the hitting time of 0, i.e.

$$\tau = \min_t \{W_t = 0\}.$$

LEMMA 5.16. *We have*

(5.23) $$\mathbf{P}(\tau = \infty) = 2\epsilon - \frac{8}{3}\epsilon^2 + O(\epsilon^3),$$

and there exists constant $C, c > 0$ such that for all $T \geq \epsilon^{-2}$ we have

(5.24) $$\mathbf{P}(T \leq \tau < \infty) \leq C\Big(\epsilon^{-2} T^{-3/2} e^{-\frac{(\epsilon^2 - c\epsilon^3)T}{2}}\Big).$$

We say that t_0 is a *record minimum* of $\{W_t\}$ if $W_t > W_{t_0}$ for all $t < t_0$.

LEMMA 5.17. *Denote by Z^w the number of record minima of W_t. Then*

$$\mathbb{E}Z^w = \frac{\epsilon^{-1}}{2} + O(1), \quad \text{and} \quad \mathbb{E}(Z^w)^2 = O(\epsilon^{-2}).$$

LEMMA 5.18. *Denote by γ the random variable*

$$\gamma = \max\{t : t \text{ is a record minimum of } W_t\}.$$

Then we have

$$\mathbb{E}\gamma = O(\epsilon^{-2}).$$

For the subcritical case we have the following.

LEMMA 5.19. *Assume $\epsilon < 0$ in the previous setting. There exists constant $C_1, C_2, c_1, c_2 > 0$ such that for all $T \geq \epsilon^{-2}$ we have*

$$\mathbf{P}(\tau \geq T) \leq C_1\Big(\epsilon^{-2} T^{-3/2} e^{-\frac{(\epsilon^2 - c_1 \epsilon^3)T}{2}}\Big),$$

and

$$\mathbf{P}(\tau \geq T) \geq c_2\Big(\epsilon^{-2} T^{-3/2} e^{-\frac{(\epsilon^2 + C_2 \epsilon^3)T}{2}}\Big).$$

Furthermore, for any fixed $k \geq 1$

$$\mathbb{E}\tau^k = O(\epsilon^{-2k+1}).$$

The proof of Lemma 5.19 can be found in [**24**] Lemma 4.

For the proof of Lemma 5.16 we will use the following proposition due to Spitzer (see [**34**]).

PROPOSITION 5.20. *Let $a_0, \ldots, a_{k-1} \in \mathbb{Z}$ satisfy $\sum_{i=0}^{k-1} a_i = -1$. Then there is precisely one $j \in \{0, \ldots, k-1\}$ such that for all $r \in \{0, \ldots, k-2\}$*

$$\sum_{i=0}^{r} a_{(j+i) \bmod k} \geq 0.$$

Proof of Lemma 5.16. Let β be a random variable distributed as $\text{Bin}(m, p)$ and let $f(s) = \mathbb{E}s^\beta$. It is a classical fact (see [**2**]) that $1 - \mathbf{P}(\tau = \infty)$ is the unique fixed point of $f(s)$ in $(0, 1)$. For $s \in (0, 1)$ we have

$$\mathbb{E}s^\beta = \big[1 - p(1-s)\big]^m = 1 - (1+\epsilon)(1-s) + \frac{(1+\epsilon)^2(1-s)^2}{2} - \frac{(1+\epsilon)^3(1-s)^3}{6} + O\big((1-s)^4\big),$$

since $(1-x)^m = 1 - mx + \frac{m^2 x^2}{2} - \frac{m^3 x^3}{6} + O(m^4 x^4)$. Write $q = 1 - s$ and put $\mathbb{E}s^\beta = s$. We get that

$$1 - (1+\epsilon)q + \frac{(1+2\epsilon)q^2}{2} - \frac{q^3}{6} + O(q^4) + O(\epsilon q^3) + O(\epsilon^2 q^2) = 1 - q.$$

Solving this gives that that $q = 2\epsilon - \frac{8}{3}\epsilon^2 + O(\epsilon^3)$, as required.

We now turn to proving (5.24). By Proposition 5.20, $\mathbf{P}(\tau = t) = \frac{1}{t}\mathbf{P}(W_t = 0)$. Since $\sum_{j=1}^t \beta_j$ is distributed as a $\text{Bin}(mt, p)$ random variable we have

$$\mathbf{P}(W_t = 0) = \binom{mt}{t-1} p^{t-1}(1-p)^{mt-(t-1)}.$$

Replacing $t - 1$ with t in the above formula only changes it by a multiplicative constant which is always between $1/2$ and 2. A straightforward computation using Stirling's approximation gives

$$(5.25) \quad \mathbf{P}(W_t = 0) = \Theta\Big\{t^{-1/2}(1+\epsilon)^t\Big(1 + \frac{1}{m-1}\Big)^{t(m-1)}\Big(1 - \frac{1+\epsilon}{m}\Big)^{t(m-1)}\Big\}.$$

Denote $x = (1+\epsilon)\big(1 + \frac{1}{m-1}\big)^{m-1}\big(1 - \frac{1+\epsilon}{m}\big)^{m-1}$, then

$$\mathbf{P}(\tau \geq T) = \sum_{t \geq T} \mathbf{P}(\tau = t) = \sum_{t \geq T} \frac{1}{t}\mathbf{P}(W_t = 0) = \Theta\Big(\sum_{t \geq T} t^{-3/2} x^t\Big).$$

This sum can be bounded above by

$$T^{-3/2} \sum_{t \geq T} x^t = T^{-3/2} \frac{x^T}{1-x}.$$

Observe that as $m \to \infty$ we have that x tends to $(1+\epsilon)e^{-\epsilon}$. By expanding $e^{-\epsilon}$ we find that

$$x = (1+\epsilon)(1 - \epsilon + \frac{\epsilon^2}{2}) + \Theta(\epsilon^3) = 1 - \frac{\epsilon^2}{2} + \Theta(\epsilon^3).$$

Using this and the previous bounds on $\mathbf{P}(\tau = \infty)$ we conclude the proof of (5.24). □

Proof of Lemma 5.17. This follows immediately since Z^w is a geometric random variable with success probability $p = P(\tau = \infty) = 2\epsilon - \frac{8}{3}\epsilon^2 + O(\epsilon^3)$ by (5.23) of Lemma 5.16. □

Proof of Lemma 5.18. At each record minimum the process has probability $\Theta(\epsilon)$ of never going below its current location by (5.23) of Lemma 5.16. It is a classical fact that the expected size of each excursion between record minimum, on the event that it is finite, is $O(\epsilon^{-1})$. Thus, by Wald's Lemma

$$\mathbb{E}(\gamma) \leq C\epsilon^{-1} \mathbb{E} Z^w = O(\epsilon^{-2}).$$

□

5.3.2. Exploration process estimates. In this section we study the process Y_t defined in Section 5.1 and provide some useful estimates.

LEMMA 5.21. *For $p = \frac{1+\epsilon}{m}$ we have*

$$\mathbf{P}\left(Y_t \geq -45\epsilon^2 m \text{ for all } 1 \leq t \leq 3\epsilon m\right) \geq 1 - 5e^{-48\epsilon^3 m}.$$

Proof. Denote by γ the stopping time

$$\gamma = \min\{t : N_t \leq m - 15\epsilon m\},$$

and consider the process $\{W_t\}$ which has i.i.d. increments distributed as $\text{Bin}(m - 15\epsilon m, p) - 1$ and $W_0 = 1$. Then we can couple the processes $\{Y_t\}$ and $\{W_t\}$ such that $Y_{t \wedge \gamma} \geq W_{t \wedge \gamma}$ and hence on the event $\gamma > 3\epsilon m$ we have

(5.26) $$\min_{t \leq 3\epsilon m} Y_t \geq \min_{t \leq 3\epsilon m} W_t.$$

Note that the expectation of the increment of W_t is $-15\epsilon - 15\epsilon^2$, thus for any positive $\alpha > 0$ the process $-\alpha W_t$ is a submartingale whence $\exp(-\alpha W_t)$ is a submartingale as well. We put $\alpha = 8\epsilon$ and applying Doob's maximal L^2 inequality (see [**10**]) yields that

$$\mathbb{E}\left[\max_{t \leq 3\epsilon m} e^{-16\epsilon W_t}\right] \leq 4 \mathbb{E}\left[e^{-16\epsilon W_{3\epsilon m}}\right].$$

Since $W_{3\epsilon m}$ is distributed as $\text{Bin}(3\epsilon(1-15\epsilon)m^2, p) - 3\epsilon m + 1$ we obtain by direct computation that

$$\mathbb{E}\left[\max_{t \leq 3\epsilon m} e^{-16\epsilon W_t}\right] \leq 4 e^{672\epsilon^3 m}.$$

Markov's inequality implies that

$$\mathbf{P}\Big(\exists t \leq 3\epsilon m \text{ with } W_t \leq -45\epsilon^2 m\Big) \leq \mathbf{P}\Big(\max_{t \leq 3\epsilon m} e^{-16\epsilon W_t} \geq e^{720\epsilon^3 m}\Big) \leq 4e^{-48\epsilon^3 m}.$$

Note that if there exists $t \leq 3\epsilon m$ with $Y_t \leq -45\epsilon^2 m$ then by (5.26) either $\gamma \leq 3\epsilon m$ or there exists $t \leq 3\epsilon m$ such that $W_t \leq -45\epsilon^2 m$. Lemma 5.2 shows that $\mathbf{P}(\gamma \leq 3\epsilon m) \leq e^{-c\epsilon m} = o(e^{-48\epsilon^3 m})$ and this concludes the proof of the lemma. □

We now use the estimates of the previous lemma to amplify Lemma 5.2.

LEMMA 5.22. *For $p = \frac{1+\epsilon}{m}$ there exists some fixed $c > 0$ such that*

$$\mathbf{P}\Big(\exists t \leq 3\epsilon m \text{ with } N_t \leq m - t - 50\epsilon^2 m\Big) \leq 9e^{-c\epsilon^3 m}.$$

Proof. Let α_i be independent random variables distributed as $\text{Bin}(m,p)$ and we couple such that $\eta_i \leq \alpha_i$ for all i. By (5.1) and the fact that Z_t is non-decreasing we have that for $t \leq 3\epsilon m$

$$(5.27) \qquad N_t \geq m - 1 - \sum_{i=1}^{t} \alpha_i - Z_{3\epsilon m}.$$

Observe that if for some positive k we have $Y_t \geq -k$ for all $t \leq T$ then $Z_T \leq k$. Thus, Lemma 5.21 together with the fact that $\{Z_t\}$ is increasing implies that

$$\mathbf{P}\Big(Z_{3\epsilon m} \geq 45\epsilon^2 m\Big) \leq 5e^{-48\epsilon^3 m}.$$

We have that $\sum_{i=1}^{t} \alpha_i$ is distributed as $\text{Bin}(mt, p)$ and has mean $t + \epsilon t$. The same argument using Doob's maximal inequality, as in the proof of Lemma 5.21, gives that

$$\mathbf{P}\Big(\exists t \leq 3\epsilon m \text{ with } \sum_{i=1}^{t} \alpha_i \geq t + 4\epsilon^2 m\Big) \leq 4e^{-c\epsilon^3 m},$$

for some fixed $c > 0$. The assertion of the lemma follows by putting the last two inequalities into (5.27). □

LEMMA 5.23. *Assume that $p = \frac{1+\epsilon}{m}$ and that $\epsilon^3 m \geq 1$. Then there exist a constant $c > 0$ such that for any a satisfying $1 \leq a \leq \sqrt{\epsilon^3 m}$ we have*

$$\mathbf{P}\Big(Y_t > 0 \text{ for all } a\sqrt{m/\epsilon} \leq t \leq 2\epsilon m - a\sqrt{m/\epsilon}\Big) \geq 1 - 2e^{-ca^2}.$$

Proof. Denote by γ the stopping time

$$\gamma = \min\{t \,:\, N_t < m - t - 50\epsilon^2 m\}.$$

Lemma 5.22 states that

$$\mathbf{P}(\gamma \leq 3\epsilon m) \leq 9e^{-c\epsilon^3 m},$$

for some constant $c > 0$. Let $\{W_t\}$ be a process with independent increments distributed as $\text{Bin}(m - t - 50\epsilon^2 m, p) - 1$ (note that the increments are not identically distributed) and $W_0 = 1$. As usual we can couple such that $Y_{t \wedge \gamma} \geq W_{t \wedge \gamma}$ for all t. Hence, if $\gamma \geq 2\epsilon m$ and there exists $t \leq 2\epsilon m$ with $Y_t \leq 0$ then it must be that $W_t \leq 0$. We conclude that it suffices to show the assertion of the lemma to the process $\{W_t\}$ and this is our next goal.

For any $\alpha > 0$ we have
$$\mathbb{E}\left[e^{-\alpha(W_t - W_{t-1})} \mid W_{t-1}\right] = e^{\alpha}[1 - p(1 - e^{-\alpha})]^{m-t-50\epsilon^2 m}.$$

We use $1 - x \leq e^{-x}$ with $x = p(1 - e^{-\alpha})$ and $1 - e^{-\alpha} \geq \alpha - \alpha^2$ for α small enough (we will eventually take $\alpha = O(\epsilon)$) to get

(5.28) $\quad \mathbb{E}\left[e^{-\alpha(W_t - W_{t-1})} \mid W_{t-1}\right] \leq e^{\alpha^2(1+\epsilon) - \alpha(\epsilon - \frac{t}{m}(1+\epsilon) - 50\epsilon^2(1+\epsilon))}.$

Thus, we learn that the process
$$e^{-\alpha W_t} e^{-(1+\epsilon)\alpha^2 t - (1+\epsilon)\alpha \frac{t^2}{2m} + \epsilon\alpha t(1 - 50\epsilon(1+\epsilon))},$$
is a supermartinagle. Write
$$f(t) = t\left[-(1+\epsilon)\alpha^2 + \epsilon\alpha(1 - 50\epsilon(1+\epsilon))\right] - t^2 \frac{(1+\epsilon)\alpha}{2m}.$$

We apply the optional stopping theorem on the stopping time $\tau = \min\{t \geq \sqrt{m/\epsilon} : W_t = 0\}$ and get that
$$\mathbb{E} e^{f(\tau)} \leq 1.$$

Direct calculation gives that when we put $\alpha = \frac{1}{3}\epsilon$ the function f attains its minimum on the interval $[a\sqrt{m/\epsilon}, \epsilon m]$ at $\tau = a\sqrt{m/\epsilon}$ for any $a \in [1, \sqrt{\epsilon^3 m}/3]$. Hence
$$\mathbf{P}\left(a\sqrt{m/\epsilon} \leq \tau \leq \epsilon m\right) \leq \mathbf{P}\left(e^{f(\tau)} \geq e^{f(a\sqrt{m/\epsilon})}\right).$$

An immediate calculation shows that $f(a\sqrt{m/\epsilon}) \geq ca\sqrt{m\epsilon^3}$ and we learn by Markov's inequality that

(5.29) $\quad \mathbf{P}\left(a\sqrt{m/\epsilon} \leq \tau \leq \epsilon m\right) \leq e^{-ca\sqrt{m\epsilon^3}} \leq e^{-ca^2},$

since $a \leq \sqrt{m\epsilon^3}$.

We are left to estimate $\mathbf{P}(\epsilon m \leq \tau \leq 2\epsilon m - a\sqrt{m/\epsilon})$. To that aim we define a new process $\{X_t\}_{t \geq 0}$ by $X_t = W_{\epsilon m + t}$. By (5.28), for positive α we have that
$$\mathbb{E}\left[e^{-\alpha(X_t - X_{t-1})} \mid X_{t-1}\right] \leq e^{\alpha^2(1+\epsilon) - \alpha(\epsilon - \frac{t + \epsilon m}{m}(1+\epsilon) - 50\epsilon^2(1+\epsilon))}.$$

This together with a straight forward computation yields that the process
$$e^{-\alpha X_t} e^{-\alpha^2 t(1+\epsilon) - \alpha\left(\frac{(1+\epsilon)t^2}{2m} + 55\epsilon^2 t\right)},$$
is a supermartingale. Write τ for the stopping time
$$\tau = \min\{t \geq 0 : X_t = 0\}.$$

Optional stopping yields that

(5.30) $\quad \mathbb{E}\left[e^{-\alpha^2 \tau(1+\epsilon) - \alpha\left(\frac{\tau^2}{2m} + 55\epsilon^2(\tau \wedge 4\epsilon m)\right)}\right] \leq \mathbb{E}\left[e^{-\alpha X_0}\right] \leq e^{\alpha^2 \epsilon m(1+\epsilon) - \alpha\left(\frac{\epsilon^2 m}{2} - 55\epsilon^3 m\right)},$

where the last inequality is an immediate calculation with (5.28) and the fact that $X_0 = W_{\epsilon m}$. Observe that the exponent on the left hand side of the previous display is
$$f(\tau) = -\alpha^2 \tau(1+\epsilon) - \alpha(\tau^2/2m + 55\epsilon^2 \tau),$$

which is a non-increasing function of τ on $[0,\infty)$. Hence, for any $a \in [1, \sqrt{\epsilon^3 m}]$ we get that

(5.31) $\quad \mathbf{P}\bigl(\tau \leq \epsilon m - a\sqrt{m/\epsilon}\bigr) \leq \mathbf{P}\bigl(e^{f(\tau)} \geq e^{f(\epsilon m - a\sqrt{m/\epsilon})}\bigr).$

We have that

$$f(\epsilon m - a\sqrt{m/\epsilon}) \geq -2\alpha^2 \epsilon m - \alpha\Bigl(\frac{\epsilon^2 m}{2} - a\sqrt{\epsilon m} + \frac{1}{2}a^2 \epsilon^{-1} + 55\epsilon^3 m\Bigr).$$

We use Markov inequality and (5.30) to get

$$\mathbf{P}\bigl(e^{f(\tau)} \geq e^{f(\epsilon m - a\sqrt{m/\epsilon})}\bigr) \leq e^{4\alpha^2 \epsilon m - \alpha\bigl(a\sqrt{\epsilon m} - \frac{1}{2}a^2 \epsilon^{-1} - 110\epsilon^3 m\bigr)}$$
$$\leq e^{4\alpha^2 \epsilon m - c\alpha a\sqrt{\epsilon m}},$$

where in the last inequality we used our assumption on a and ϵ. We choose $\alpha \approx a(\epsilon m)^{-1/2}$ that minimizes the last expression. This yields

$$\mathbf{P}\bigl(e^{f(\tau)} \geq e^{f(\epsilon m - a\sqrt{m/\epsilon})}\bigr) \leq e^{-ca^2}.$$

We put this into (5.31), which together with (5.29) yields the assertion of the lemma. $\square$

LEMMA 5.24. *Assume that* $p = \frac{1+\epsilon}{m}$. *Write* $\tau = \min\{t : Y_t = 0\}$, *then for any small* $\alpha > 0$

$$\mathbb{E}\bigl[e^{\alpha Y_{\epsilon^{-2}}} \mid \tau \geq \epsilon^{-2}\bigr] \leq C e^{2\alpha \epsilon^{-1} + \alpha^2 \epsilon^{-2}}.$$

Proof. We have that $\mathbf{P}(\tau \geq \epsilon^{-2}) \geq c\epsilon$. To see this we perform the usual argument of bounding Y_t below by a process of independent increments (until a stopping time, using Lemma 5.2) and using Lemma 5.16. This has been done in this section several times so we omit the details. Thus, it suffices to bound from above $\mathbb{E} e^{\alpha Y_{\epsilon^{-2}}} \mathbf{1}_{\{\tau \geq \epsilon^{-2}\}}$. Since we can bound Y_t by a process W_t which has i.i.d. $\text{Bin}(m, p) - 1$ increments, it suffices to bound the same expectation for W_t. Write $\gamma = \min\{t : W_t = 0 \text{ or } W_t \geq \epsilon^{-1}\}$. We have

$$\mathbb{E} e^{\alpha W_{\epsilon^{-2}}} \mathbf{1}_{\{\tau \geq \epsilon^{-2}\}} \leq \mathbb{E} e^{\alpha W_{\epsilon^{-2}}} \mathbf{1}_{\{\tau \geq \epsilon^{-2}, \gamma \geq \epsilon^{-2}\}} + \mathbb{E} e^{\alpha W_{\epsilon^{-2}}} \mathbf{1}_{\{\tau \geq \epsilon^{-2}, \gamma < \epsilon^{-2}\}}.$$

For the first term on the right hand side we note that on $\gamma \geq \epsilon^{-2}$ we have that $W_{\epsilon^{-2}} \leq \epsilon^{-1}$, so

$$\mathbb{E} e^{\alpha W_{\epsilon^{-2}}} \mathbf{1}_{\{\tau \geq \epsilon^{-2}, \gamma \geq \epsilon^{-2}\}} \leq C\epsilon e^{\alpha \epsilon^{-1}}.$$

For the second term we condition on $\{\tau \geq \epsilon^{-2}, \gamma < \epsilon^{-2}\}$ (which implies $W_\gamma \geq \epsilon^{-1}$ and $\gamma < \epsilon^2$) to get that
(5.32)
$$\mathbb{E} e^{\alpha W_{\epsilon^{-2}}} \mathbf{1}_{\{\tau \geq \epsilon^{-2}, \gamma < \epsilon^{-2}\}} \leq \mathbf{P}(W_\gamma \geq \epsilon^{-1}) \mathbb{E}[e^{\alpha W_\gamma} e^{\alpha(W_{\epsilon^{-2}} - W_\gamma)} \mid W_\gamma \geq \epsilon^{-1}, \gamma < \epsilon^{-2}].$$

We have that $\mathbf{P}(W_\gamma \geq \epsilon^{-1}) = O(\epsilon)$ by Lemma 7 of [26]. We condition in addition on W_γ and γ and pull out the $e^{\alpha W_\gamma}$ factor. By the strong Markov property we have that conditioned on all these, the random variable $W_{\epsilon^{-2}} - W_\gamma$ is distributed as the sum of $\epsilon^{-2} - \gamma$ i.i.d. copies of $\text{Bin}(m, p) - 1$ random variables. Thus,

$$\mathbb{E}[e^{\alpha(W_{\epsilon^{-2}} - W_\gamma)} \mid W_\gamma, \gamma < \epsilon^2] \leq e^{-\alpha \epsilon^{-2}} [1 + p(e^\alpha - 1)]^{m\epsilon^{-2}}.$$

Furthermore, Lemma 5 of [**25**] states that conditioned on $W_\gamma \geq \epsilon^{-1}$ and $\gamma < \epsilon^2$ the distribution of $W_\gamma - \epsilon^{-1}$ is bounded above by $\mathrm{Bin}(m,p)$, whence
$$\mathbb{E}[e^{\alpha W_\gamma} \mid W_\gamma \geq \epsilon^{-1}, \gamma < \epsilon^{-2}] \leq e^{\alpha \epsilon^{-1}}[1 + p(e^\alpha - 1)]^m \,.$$
Putting this back into (5.32) gives
$$\mathbb{E}e^{\alpha W_{\epsilon^{-2}}} \mathbf{1}_{\{\tau \geq \epsilon^{-2}, \gamma < \epsilon^{-2}\}} \leq C\epsilon e^{-\alpha(\epsilon^{-2} - \epsilon^{-1})}[1 + p(e^\alpha - 1)]^{m(\epsilon^{-2}+1)} \,.$$
Putting all these together we get
$$\begin{aligned}
\mathbb{E}\bigl[e^{\alpha Y_{\epsilon^{-2}}} \mid \tau \geq \epsilon^{-2}\bigr] &\leq C e^{\alpha \epsilon^{-1}} + C e^{-\alpha(\epsilon^{-2}-\epsilon^{-1})}[1 + p(e^\alpha-1)]^{m(\epsilon^{-2}+1)} \\
&\leq C e^{\alpha \epsilon^{-1}} + C e^{-\alpha(\epsilon^{-2}-\epsilon^{-1})} e^{(1+\epsilon)(\alpha+\alpha^2)(\epsilon^{-2}+1)} ,
\end{aligned}$$
The lemma follows now by an immediate calculation. $\square$

LEMMA 5.25. *Let $p = \frac{1+\epsilon}{m}$ and assume $\epsilon^3 m \geq 1$. Then for any $\ell > 0$, we have*
$$\mathbf{P}(|\mathcal{C}(v)| \geq 2\epsilon m + \ell) \leq C\epsilon e^{\frac{-c\ell^2(2\epsilon m + \ell)}{m^2}} \,.$$

Proof. We assume that $\ell \geq 2\sqrt{m/\epsilon}$ since otherwise the exponential is of constant order and the assertion of the lemma follows simply from Lemma 5.16. Recall that $|\mathcal{C}(v)|$ is distributed as the first hitting time τ of Y_t at 0. We put $T = 2\epsilon m + \ell$ and condition on $Y_{\epsilon^{-2}}$ and on $\tau \geq \epsilon^{-2}$. That is,

(5.33) $\quad \mathbf{P}(\tau \geq 2\epsilon m + \ell) = \mathbf{P}(\tau \geq \epsilon^{-2}) \mathbb{E}\bigl[\mathbf{P}(\tau \geq T \mid Y_{\epsilon^{-2}}, \tau \geq \epsilon^{-2})\bigr] \,.$

Since Y_t is bounded above by a process with increments distributed as $\mathrm{Bin}(m,p)-1$, we learn by Lemma 5.16 that $\mathbf{P}(\tau \geq \epsilon^{-2}) = O(\epsilon)$. The second term will give us the exponential in the assertion of the Lemma simply because Y_T has small probability of being positive at this time. Indeed, since the increments of Y_t are stochastically bounded above by $\mathrm{Bin}(m-t,p)-1$ we have that for any small $\alpha > 0$
$$\mathbb{E}\Bigl[e^{\alpha(Y_t - Y_{t-1})} \mid Y_{t-1}\Bigr] \leq e^{-\alpha}[1 + p(e^\alpha - 1)]^{m-t} \leq e^{-\alpha + (1+\epsilon)(\alpha+\alpha^2)(1-t/m)} ,$$
since $e^\alpha - 1 \leq \alpha + \alpha^2$ for small enough α. Summing this over t ranging from ϵ^{-2} to T gives
$$\begin{aligned}
\mathbb{E}\bigl[e^{\alpha Y_T} \mid Y_{\epsilon^{-2}}, \tau \geq \epsilon^{-2}\bigr] &\leq e^{-\alpha(T - \epsilon^{-2}) + (1+\epsilon)(\alpha+\alpha^2)\left(T - \epsilon^{-2} - \frac{T^2 - \epsilon^{-4}}{2m}\right)} e^{\alpha Y_{\epsilon^{-2}}} \\
&\leq e^{\alpha^2 T (1+\epsilon) - \alpha\left[\frac{T^2 - \epsilon^{-4}}{2m} - \epsilon T\right]} e^{\alpha Y_{\epsilon^{-2}}} \,.
\end{aligned}$$
Hence,
$$\begin{aligned}
\mathbb{E}\bigl[e^{\alpha Y_T} \mid \tau \geq \epsilon^{-2}\bigr] &\leq \mathbb{E}\bigl[e^{\alpha Y_{\epsilon^{-2}}} \mid \tau \geq \epsilon^{-2}\bigr] e^{\alpha^2 T(1+\epsilon) - \alpha\left[\frac{T^2 - \epsilon^{-4}}{2m} - \epsilon T\right]} \\
&\leq C e^{\alpha^2 (T + \epsilon^{-2})(1+\epsilon) - \alpha\left[\frac{T^2-\epsilon^{-4}}{2m} - \epsilon T - 2\epsilon^{-1}\right]} ,
\end{aligned}$$
where the last inequality is due to Lemma 5.24. Hence, by Markov's inequality this is also an upper bound on $\mathbf{P}(Y_T \geq 0 \mid \tau \geq \epsilon^{-2})$ which is what we aim to estimate. We now choose α
$$\alpha = \frac{\frac{T^2 - \epsilon^{-4}}{2m} - \epsilon T - 2\epsilon^{-1}}{2(T + \epsilon^{-2})} ,$$

which is positive and of order ℓ/m since $\ell \geq 2\sqrt{m/\epsilon}$ and minimizes the above expectation. We get that

$$\mathbf{P}(Y_T \geq 0 \mid \tau \geq \epsilon^{-2}) \leq Ce^{-\frac{cT(T-2\epsilon m)^2}{m^2}},$$

for some $c > 0$ by a straightforward calculation, concluding our proof. $\square$

5.3.3. Proof of near-critical random graph theorems. We are now ready to prove the Theorems stated in Section 5.3.

Proof of Theorem 5.8. We begin by proving the upper bound on $\mathbb{E}|\mathcal{C}_1|$. For any positive integer ℓ define by X_ℓ the random variable

$$X_\ell = \left|\left\{v \,:\, |\mathcal{C}(v)| \geq \ell\right\}\right|.$$

Observe that if $|\mathcal{C}_1| \geq \ell$, then we must have that $|X_\ell| \geq |\mathcal{C}_1|$. Thus for any positive integer ℓ we have

(5.34) $$\mathbb{E}|\mathcal{C}_1| \leq \ell\,\mathbf{P}(|\mathcal{C}_1| < \ell) + \mathbb{E}X_\ell.$$

We take $\ell = \frac{1}{20}\epsilon m$ and since Lemma 5.23 implies that $\mathbf{P}(|\mathcal{C}_1| \leq \ell) \leq Ce^{-c\epsilon^3 m}$ and $\epsilon^3 m \geq A\log m$ we have that the first term on the right hand side of (5.34) is $o(1)$. We now turn to bound the second term on the right hand side of (5.34). Since $\mathbb{E}X_\ell = m\mathbf{P}(|\mathcal{C}(v)| \geq \ell)$ it suffices to bound from above $\mathbf{P}(|\mathcal{C}(v)| \geq \ell)$. Recall that $|\mathcal{C}(v)|$ is the hitting time of the process $\{Y_t\}$ at 0. Let $\{W_t\}$ be a process with independent increments distributed as $\text{Bin}(m,p) - 1$ and $W_0 = 1$, as in Lemma 5.16. Let $\tau = \min_t\{W_t = 0\}$ be the hitting time of W_t at 0, then it is clear that we can couple W_t and Y_t such that $|\mathcal{C}(v)| \leq \tau$. Thus

$$\mathbf{P}(|\mathcal{C}(v)| \geq \ell) \leq \mathbf{P}(\tau \geq \ell) = \mathbf{P}(\tau = \infty) + \mathbf{P}(\ell \leq \tau < \infty).$$

We now apply Lemma 5.16 with $T = \ell = \frac{1}{20}\epsilon m$ and get by the previous display that

$$\begin{aligned}\mathbf{P}(|\mathcal{C}(v)| \geq \epsilon m/20) &\leq 2\epsilon - \frac{8}{3}\epsilon^2 + O(\epsilon^3) + C_1\epsilon^{-7/2}m^{-3/2}e^{-\epsilon^3 m/4} \\ &= 2\epsilon - \frac{8}{3}\epsilon^2 + O(\epsilon^3),\end{aligned}$$

as long as $\epsilon^3 m \geq A\log m$ for large enough A. We conclude that

$$\mathbb{E}X_\ell \leq 2\epsilon m - \frac{8}{3}\epsilon^2 m + O(\epsilon^3 m),$$

which together with (5.34) concludes the proof of the upper bound on $\mathbb{E}|\mathcal{C}_1|$.

We turn to the proof of the lower bound on $\mathbb{E}|\mathcal{C}_1|$. Recall that at each record minimum of the process $\{Y_t\}$ we are starting the exploration of a new component. Write

$$\gamma = \max\left\{t \leq \epsilon m \,:\, Y_t \text{ is at a record minimum}\right\},$$

and

$$\tau = \min\{t \geq 0 : Y_{\epsilon m + t} < 0\}.$$

Then we have that

(5.35) $$|\mathcal{C}_1| \geq \epsilon m - \gamma + \tau.$$

Thus, in order to complete the proof we will provide an upper bound on $\mathbb{E}\gamma$ and a lower bound on $\mathbb{E}\tau$. Let $\{W_t\}$ be a process defined as in Lemma 5.18 with i.i.d. increments distributed as $\text{Bin}(m(1-\epsilon/2), p) - 1$. Define the stopping time β by

$$\beta = \min\{t : N_t \leq m(1-\epsilon/2)\},$$

then it is clear we can couple $\{Y_{t\wedge\beta}\}$ with $\{W_{t\wedge\beta}\}$ such that the increments of the first are larger than of the latter process. This guarantees that every record minimum of the first process is also a record minimum of the second, and thus if we put

$$\gamma^w = \max\left\{t : W_t \text{ is at a record minimum}\right\},$$

then we can couple such that

$$\gamma \mathbf{1}_{\{\text{no record minima at times } [\epsilon m/10, \epsilon m]\}} \leq \gamma^w + \epsilon m \mathbf{1}_{\{\beta \leq \epsilon m/10\}}.$$

Lemma 5.23 shows that the probability that there is a record minimum at some time between $\epsilon m/10$ and ϵm decays faster than m^{-2} provided that $\epsilon^3 m \geq A \log m$ for A large enough. Hence, taking expectations on both sides and using Lemma 5.18 and Lemma 5.2 gives that $\mathbb{E}\gamma = O(\epsilon^{-2})$.

We now turn to give a lower bound on $\mathbb{E}\tau$. We begin by estimating $\mathbb{E}\tau^2$. As before, define the process $\{X_t\}_{t\geq 0}$ by $X_t = Y_{\epsilon m+t}$ and note that $X_t - X_{t-1} = \eta_{\epsilon m+t}$. For any t such that $N_{t+\epsilon m} \geq m - (t+\epsilon m) - 50\epsilon^2 m$ we have

$$\mathbb{E}[X_{t+1} - X_t \mid \mathcal{F}_t] \geq -\frac{t(1+\epsilon)}{m} - 55\epsilon^2.$$

Thus the process $\{X_{t\wedge T} + \frac{(t\wedge T)^2(1+\epsilon)}{2m} + 55\epsilon^2(t\wedge T)\}$ is a submartingale, where T is defined as

$$T = \min\{t - \epsilon m : t \geq \epsilon m, N_t \leq m - t - 50\epsilon^2 m\}.$$

Optional stopping yields that

(5.36) $$\mathbb{E}(\tau \wedge T)^2 \geq \frac{2m}{1+\epsilon}(\mathbb{E}X_0 - \mathbb{E}X_{\tau\wedge T}) - \frac{110\epsilon^2 m}{1+\epsilon}\mathbb{E}[\tau \wedge T].$$

By Lemma 5.23, we have

$$\mathbf{P}(\tau < \epsilon m - a\sqrt{m/\epsilon}) \leq e^{-ca^2}.$$

Also by lemma 5.23, one can deduce

$$\begin{aligned}\mathbf{P}(\tau > \epsilon m + a\sqrt{m/\epsilon}) &\leq \mathbf{P}(\tau > \epsilon m + a\sqrt{m/\epsilon}, Y_t > 0 \text{ for } t \in [\tfrac{a}{2}\sqrt{m/\epsilon}, \epsilon m]) + e^{-ca^2} \\ &\leq \mathbf{P}(|\mathcal{C}_1| > 2\epsilon m + \tfrac{a}{2}\sqrt{m/\epsilon}) + e^{-ca^2} \\ &\leq \mathbf{P}(X_{2\epsilon m a\sqrt{m/\epsilon}/2} > 2\epsilon m) + e^{-ca^2},\end{aligned}$$

where $X_{2\epsilon m a\sqrt{m/\epsilon}/2}$ is the number of vertices v such that $|\mathcal{C}_v| \geq 2\epsilon m + \tfrac{a}{2}\sqrt{m/\epsilon}$ as defined in the beginning of the proof. By Lemma 5.25, we have

$$\mathbb{E}X_{2\epsilon m a\sqrt{m/\epsilon}/2} \leq Cm\epsilon e^{-ca^2}.$$

Plugging this into the previous inequality and using Markov's inequality shows that $\mathbb{E}[\tau \wedge T] = O(\epsilon m)$ and

(5.37) $$\mathbf{P}(|\tau - \epsilon m| > a\sqrt{m/\epsilon}) \leq Ce^{-ca^2}.$$

Lemma 5.22 shows that $\mathbf{P}(T \leq 2\epsilon m) \leq m^{-2}$, and so $\mathbb{E}X_{\tau \wedge T} = o(1)$ and $\mathbb{E}(\tau \wedge T)^2 = \mathbb{E}\tau^2 + o(1)$. We get that
$$\mathbb{E}\tau^2 \geq 2m\mathbb{E}Y_{\epsilon m} - o(1).$$
We bound from below $\mathbb{E}Y_{\epsilon m}$ using the approximating process $\widetilde{Y}_t$ defined in (5.14). We have that $\mathbb{E}\widetilde{Y}_{\epsilon m} = \epsilon^2 m^2 + O(\epsilon^3 m)$ and using (5.15) and Lemma 5.26 we deduce the same estimate for $\mathbb{E}Y_{\epsilon m}$. This yields that
$$\mathbb{E}\tau^2 \geq \epsilon^2 m^2 - C\epsilon^3 m^2,$$
for some $C > 0$. Inequality (5.37) gives that for some $C > 0$ we have
$$\mathrm{Var}(\tau) \leq \mathbb{E}\Big[(\tau - \epsilon m)^2\Big] \leq \frac{Cm}{\epsilon}.$$
We conclude
$$\mathbb{E}\tau = \sqrt{\mathbb{E}\tau^2 - \mathrm{Var}(\tau)} \geq \epsilon m \sqrt{1 - C\epsilon - \frac{C}{\epsilon^3 m}} \geq \epsilon m - C\epsilon^2 m - C\epsilon^{-2},$$
since $\sqrt{1-x} \geq 1 - x$ for $x \in (0, 1)$. Using this and our estimate on $\mathbb{E}\gamma$ in (5.35) finishes the proof. $\square$

Proof of Theorem 5.9. Since component sizes are excursions' length above past minima and $Y_0 = 1$, Lemma 5.23 immediately yields the bound
$$(5.38) \qquad \mathbf{P}\Big(|\mathcal{C}_1| \leq 2\epsilon m - A\sqrt{\frac{m}{\epsilon}}\Big) \leq e^{-cA^2},$$
valid for any A satisfying $1 \leq A \leq \sqrt{\epsilon^3 m}$. For the upper bound we use Lemma 5.25 stating that
$$\mathbf{P}(|\mathcal{C}(v)| \geq 2\epsilon m + A\sqrt{m/\epsilon}) = O(\epsilon e^{-cA^2}).$$
Write $X = |\{v : |\mathcal{C}(v)| \geq 2\epsilon m + A\sqrt{m/\epsilon}\}|$ so that $\mathbb{E}X = O(\epsilon m e^{-cA^2})$. As usual we have
$$\mathbf{P}(|\mathcal{C}_1| \geq 2\epsilon m + A\sqrt{m/\epsilon}) \leq \mathbf{P}(X \geq 2\epsilon m) = O(e^{-cA^2}),$$
by Markov's inequality, concluding the proof. $\square$

Proof of Corollary 5.10. Part (i) of the corollary follows immediately from Theorem 5.9 by integration. Indeed,
$$\mathbb{E}\Big[\big||\mathcal{C}_1| - 2\epsilon m\big|^k\Big] = \sum_\ell \ell^{k-1} \mathbf{P}(||\mathcal{C}_1| - 2\epsilon m| > \ell)$$
$$\leq \sum_{\ell=1}^{\sqrt{\frac{m}{\epsilon}}} \ell^{k-1} + C \sum_{\ell=\sqrt{\frac{m}{\epsilon}}}^{\epsilon m} \ell^{k-1} e^{\frac{-c\ell^2 \epsilon}{m}} + \sum_{\ell \geq \epsilon m} \ell^{k-1} e^{\frac{-c\ell^3}{m^2}},$$
where we bounded the second sum on the right hand side using Theorem 5.9 and the last sum using Lemma 5.25 (which is valid for all $\ell > 0$ and not limited by $\ell \leq \sqrt{\epsilon^3 m}$) and the usual Markov inequality on the variable $X = |\{v : |\mathcal{C}(v)| \geq 2\epsilon m + \ell\}|$. A quick calculation now shows each term is of order at most $(m/\epsilon)^{k/2}$, concluding our proof. $\square$

Proof of Theorem 5.12. We begin by proving (i). As before, $|\mathcal{C}(v)|$ is stochastically dominated by the random variable τ defined in Lemma 5.19. This Lemma gives that for any fixed $k \geq 1$

$$\mathbb{E}|\mathcal{C}(v)|^k = O(\epsilon^{-2k+1}).$$

Number the vertices of $G(m, p)$ arbitrarily $v_1, \ldots, v_m$ and observe that

$$\sum_{j \geq 1} |\mathcal{C}_j|^k = \sum_{i=1}^m |\mathcal{C}(v_i)|^{k-1},$$

because each component $\mathcal{C}_j$ is counted in the sum in the right hand size precisely $|\mathcal{C}_j|$ times. By symmetry we learn that

$$\mathbb{E} \sum_{j \geq 1} |\mathcal{C}_j|^k = m \mathbb{E}|\mathcal{C}(v)|^{k-1} = O(m\epsilon^{-2k+3}),$$

finishing the first assertion of the theorem.

We proceed to prove (ii). Recall that $\sum_j |\mathcal{C}_j|^2 = \sum_v |\mathcal{C}(v)|$. Thus,

$$\mathbb{E}\Big(\sum_j |\mathcal{C}_j|^2\Big)^2 = \mathbb{E}\sum_{v,w} |\mathcal{C}(v)||\mathcal{C}(w)|1_{\{\mathcal{C}(v)=\mathcal{C}(w)\}} + \mathbb{E}\sum_{v,w} |\mathcal{C}(v)||\mathcal{C}(w)|1_{\{\mathcal{C}(v)\neq \mathcal{C}(w)\}}.$$

The first term on the right hand side is $\sum_v \mathbb{E}|\mathcal{C}(v)|^3$ which equals $\mathbb{E}\sum_j |\mathcal{C}_j|^4$ and is upper bounded by $O(m\epsilon^{-5})$ by part (i) of the theorem. This bound is $O(m^2\epsilon^{-2})$ since $\epsilon^3 m \geq 1$. For the second term we note that we can write $\mathbb{E}\sum_{v,w} |\mathcal{C}(v)||\mathcal{C}(w)|1_{\{\mathcal{C}(v)\neq\mathcal{C}(w)\}}$ as

$$\mathbb{E}\sum_w |\mathcal{C}(w)| \sum_v |\mathcal{C}(v)|1_{\{v \notin \mathcal{C}(w)\}} = \mathbb{E}\sum_w |\mathcal{C}(w)| \sum_{\{v \notin \mathcal{C}(w)\}} |\mathcal{C}(v)|.$$

Conditioned on $\mathcal{C}(w)$ the distribution of the rest of the graph is also subcritical random graph with ϵ' bigger than ϵ. Thus the estimate of part (i) of the theorem (together with the fact that $\sum_v |C(v)| = \sum_j |\mathcal{C}_j|^2$) can be applied and we may bound

$$\mathbb{E}\sum_{v,w} |\mathcal{C}(v)||\mathcal{C}(w)|1_{\{\mathcal{C}(v)\neq\mathcal{C}(w)\}} \leq Cm\epsilon^{-1}\mathbb{E}\sum_w |\mathcal{C}(w)| = O(m^2\epsilon^{-2}),$$

which finishes the proof of (ii).

To prove part (iii) of the theorem, let W_t be a process with i.i.d. increment distributed as $\mathrm{Bin}(m - 5\epsilon^{-2}, \frac{1-\epsilon}{m}) - 1$ and $W_0 = 1$. Let

$$\tau = \min\{t : N_t < m - 5\epsilon^{-2}\}.$$

As usual we can couple such that $Y_{t \wedge \tau} \geq W_{t \wedge \tau}$. Let $\gamma = \min\{t : W_t \leq 0\}$. For any T We have

$$\begin{aligned}
\mathbf{P}(\gamma \geq T) &= \mathbf{P}(\gamma \geq T, \tau \leq T) + \mathbf{P}(\gamma \geq T, \tau > T) \\
&\leq \mathbf{P}(\tau \leq T) + \mathbf{P}(|\mathcal{C}(v)| \geq T),
\end{aligned}$$

which implies

(5.39) $$\mathbf{P}(|\mathcal{C}(v)| \geq T) \geq \mathbf{P}(\gamma \geq T) - \mathbf{P}(\tau \leq T).$$

Put $T = \epsilon^{-2}$ we have by Lemma 5.2 that

$$\mathbf{P}(\tau \leq T) \leq e^{-c\epsilon^{-2}}.$$

Furthermore, Lemma 5.19 shows that
$$\mathbf{P}(\gamma \geq \epsilon^{-2}) \geq c\epsilon,$$
for some constant $c > 0$. Thus, by (5.39) we get that
$$\mathbf{P}(|\mathcal{C}(v)| \geq \epsilon^{-2}) \geq c\epsilon - e^{-c\epsilon^{-2}},$$
which implies $\mathbb{E}|\mathcal{C}(v)| \geq c\epsilon^{-1}$ and concludes the proof. $\square$

Proof of Theorem 5.13. The proof of (i) is a calculation using Lemma 5.25. We have
$$\mathbb{E}|\mathcal{C}(v)|^k = \sum_{\ell=1}^{\epsilon^{-2}} \ell^{k-1} \mathbf{P}(|\mathcal{C}(v)| \geq \ell) + \sum_{\ell=\epsilon^{-2}}^{10\epsilon m} \ell^{k-1} \mathbf{P}(|\mathcal{C}(v)| \geq \ell) + \sum_{\ell=10\epsilon m}^{m} \ell^{k-1} \mathbf{P}(|\mathcal{C}(v)| \geq k).$$

For the first sum we use the estimate $\mathbf{P}(|\mathcal{C}(v)| \geq l) \leq O(\epsilon + \ell^{-1/2})$ appearing in the proof of Proposition 1 of [26]. We get
$$\sum_{\ell=1}^{\epsilon^{-2}} \ell^{k-1} \mathbf{P}(|\mathcal{C}(v)| \geq \ell) \leq C \sum_{\ell=1}^{\epsilon^{-2}} \ell^{k-1}(\epsilon + \ell^{-1/2}) = O(\epsilon^{-2k+1}).$$

For the second sum, since Y_t is bounded above by a process with i.i.d. increments Bin(m,p)-1, each term is of order ϵ by Lemma 5.16. This gives the main contribution of $O(\epsilon^{k+1} m^k)$. Lastly, the third sum we bound using Lemma 5.25 to get
$$\sum_{\ell=10\epsilon m}^{m} \ell^{k-1} \mathbf{P}(|\mathcal{C}(v)| \geq k) \leq C\epsilon \sum_{\ell=10\epsilon m}^{m} \ell^{k-1} e^{-cm^{-2}\ell^3}.$$

Since $\epsilon^3 m \geq 1$ we may bound the sum above by summing from $m^{2/3}$ to m. A straightforward calculation then gives that
$$\sum_{\ell=10\epsilon m}^{m} \ell^{k-1} \mathbf{P}(|\mathcal{C}(v)| \geq k) \leq C\epsilon(\epsilon m)^{k-1} m^{2/3} = O(\epsilon^{k+1} m^k),$$

which finishes the proof of (i). We proceed to prove (ii). We have that
$$(5.40) \quad \mathbb{E} \sum_{j\geq 2} |\mathcal{C}_j|^k = \mathbb{E} \sum_{j\geq 2} |\mathcal{C}_j|^k \mathbf{1}_{\{|\mathcal{C}_1| < 1.5\epsilon m\}} + \mathbb{E} \sum_{j\geq 2} |\mathcal{C}_j|^k \mathbf{1}_{\{|\mathcal{C}_1| \geq 1.5\epsilon m\}}.$$

For the first term of (5.40) we apply FKG inequality to get
$$\mathbb{E} \sum_{j\geq 2} |\mathcal{C}_j|^k \mathbf{1}_{\{|\mathcal{C}_1| < 1.5\epsilon m\}} \leq \mathbb{E} \sum_{j\geq 1} |\mathcal{C}_j|^k \mathbf{1}_{\{|\mathcal{C}_1| < 1.5\epsilon m\}} \leq \mathbf{P}(|\mathcal{C}_1| < 1.5\epsilon m) \mathbb{E} \sum_{j\geq 1} |\mathcal{C}_j|^k.$$

By Theorem 5.9, we have
$$\mathbf{P}(|\mathcal{C}_1| < 1.5\epsilon m) \leq Ce^{-c\epsilon^3 m},$$
and so
$$\mathbb{E} \sum_{j\geq 2} |\mathcal{C}_j|^k \mathbf{1}_{\{|\mathcal{C}_1| < 1.5\epsilon m\}} \leq Ce^{-c\epsilon^3 m} m \mathbb{E}|\mathcal{C}(v)|^{k-1}.$$

By part (i) of the theorem this is at most $C\epsilon^k m^k e^{-c\epsilon^3 m}$ which is $O(m\epsilon^{-2k+3})$ since $\epsilon \geq m^{-1/3}$. This shows the required bound for the first term of (5.40).

To take care of the second term of (5.40) we condition on $\mathcal{C}_1$ and note that the graph remaining is distributed as $G(m - |\mathcal{C}_1|, p)$ conditioned on the event of not having a component larger than $|\mathcal{C}_1|$. But since $|\mathcal{C}_1| \geq 1.5\epsilon m$ this random graph is in the subcritical regime, and the probability of having such a component is smaller than $1/2$ (in fact, it is exponentially small). The required estimate follows by part (i) of Theorem 5.12. This finishes the proof of (ii).

The proof of (iii) goes in similar lines of (ii). We have

$$\mathbb{E}(\sum_{j\geq 2}|\mathcal{C}_j|^2)^2 = \mathbb{E}(\sum_{j\geq 2}|\mathcal{C}_j|^2)^2\mathbf{1}_{\{|\mathcal{C}_1|<1.5\epsilon m\}} + \mathbb{E}(\sum_{j\geq 2}|\mathcal{C}_j|^2)^2\mathbf{1}_{\{|\mathcal{C}_1|\geq 1.5\epsilon m\}}.$$

As in the proof of (ii), to control the first term we use FKG inequality, extract $\mathbf{P}(|\mathcal{C}_1|<1.5\epsilon m)$ and bound the rest by $\mathbb{E}(\sum_{j\geq 1}|\mathcal{C}_j|^2)^2$ (instead of $j\geq 2$). The analysis performed in the proof of part (ii) of Theorem 5.12 shows that $\mathbb{E}(\sum_{j\geq 1}|\mathcal{C}_j|^2)^2$ is controlled by $(\mathbb{E}\sum_{j\geq 1}|\mathcal{C}_j|^2)^2$. We get that

$$\mathbb{E}(\sum_{j\geq 2}|\mathcal{C}_j|^2)^2\mathbf{1}_{\{|\mathcal{C}_1|<1.5\epsilon m\}} \leq Ce^{-c\epsilon^3 m}m^4\epsilon^4 = O(m^2\epsilon^{-2}).$$

To control the second term, as in the proof of (ii), we condition on $\mathcal{C}_1$ and use part (ii) of Theorem 5.12 to estimate the remaining subcritical graph. This is done identically to part (ii) and we omit the details. $\square$

Proof of Theorem 5.14. We will use a second moment method. First we show that

$$\mathbb{E}\sum_{|\mathcal{C}_j|\leq \delta\sqrt{m}}|\mathcal{C}_j|^2 \geq cm^{5/4},$$

for some $c=c(\delta)>0$. Indeed, we have
(5.41)
$$\mathbb{E}|\mathcal{C}(v)|\mathbf{1}_{\{|\mathcal{C}(v)|\leq\delta\sqrt{m}\}} \geq \mathbb{E}|\mathcal{C}(v)|\mathbf{1}_{\{\frac{\delta}{2}\sqrt{m}\leq|\mathcal{C}(v)|\leq\delta\sqrt{m}\}} \geq \frac{\delta}{2}\sqrt{m}\mathbf{P}(\frac{\delta}{2}\sqrt{m}\leq|\mathcal{C}(v)|\leq\delta\sqrt{m}).$$

We proceed further by restricting to the case that $\mathcal{C}_v$ is tree. Indeed, we have

$$\mathbf{P}(\frac{\delta}{2}\sqrt{m}\leq|\mathcal{C}(v)|\leq\delta\sqrt{m}) \geq \sum_{k=\delta/2\sqrt{m}}^{\delta\sqrt{m}}\mathbf{P}(|\mathcal{C}(v)|=k,\mathcal{C}(v)\text{ is a tree})$$

$$= \sum_{k=\delta/2\sqrt{m}}^{\delta\sqrt{m}}\binom{m-1}{k-1}k^{k-2}p^{k-1}(1-p)^{k(m-k)+\binom{k}{2}-(k-1)}.$$

A quick calculation using Stirling's formula gives that for all such k, each summand is of order $\Theta(m^{-4/3})$ and so the probability is of order at least $m^{-1/4}$ and the expectation in (5.41) is of order at least $m^{1/4}$. This gives the first moment estimate since

$$\mathbb{E}\sum_{|\mathcal{C}_j|\leq\delta\sqrt{m}}|\mathcal{C}_j|^2 = \mathbb{E}\sum_{v:|\mathcal{C}(v)|\leq\delta\sqrt{m}}|C(v)|.$$

We continue with the second moment estimate. By Theorem 5.12 the second moment satisfies

$$\mathbb{E}[\sum_j|\mathcal{C}_j|^2]^2 = O(m^{5/2}),$$

and so the assertion of the Theorem follows by the inequality (see [10])

$$\mathbf{P}(V>a) \geq \frac{(\mathbb{E}V-a)^2}{\mathbb{E}V^2},$$

valid for any non-negative random variable V and $a < \mathbb{E}V$. $\square$

Now we turn to the proof of Theorem 5.15. Recall that Z_t counts the number of record minima of $\{Y_s\}$ before time t.

LEMMA 5.26. *For any fixed $\delta \in (0, 1/10)$, there exists an universal constant $C > 0$ such that as long as $\epsilon^3 m \geq 1$ we have*

$$\mathbb{E}Z_{\delta\epsilon m} \leq \frac{1}{2[(1 - 5\delta - 5\delta\epsilon)\epsilon]} + O(1),$$

and

$$\mathbb{E}Z_{\delta\epsilon m}^2 = O(\epsilon^{-2}).$$

Proof. Define the stopping time τ by

$$\tau = \min\{t : N_t \leq m(1 - 5\delta\epsilon)\},$$

and $\{W_t\}$ to be the process with increments distributed as $\mathrm{Bin}(m(1 - 5\delta\epsilon), p)$ and $W_0 = 1$. As usual we can couple such that $Y_{t \wedge \tau} \geq W_{t \wedge \tau}$ and that the increments of the first process are always larger than of the second. This guarantees that the number of record minimum of $Y_{t \wedge \tau}$ is bounded from above by the record minimum of $W_{t \wedge \tau}$. Denote by Z^w the number of record minima of the process $\{W_t\}$, then by the above discussion we have

$$\mathbb{E}Z_{\delta\epsilon m} \leq \delta\epsilon m \mathbb{E}\mathbf{1}_{\{\tau < \delta\epsilon m\}} + \mathbb{E}Z^w.$$

The order of the first term can be arbitrarily small since $\mathbf{P}(\tau < \delta\epsilon m)$ is exponentially small in ϵm by Lemma 5.2. Lemma 5.17 bound the second term by the required amount. This concludes the bound on $\mathbb{E}Z_{\delta\epsilon m}$. For the second moment estimate, note that by the same argument, we have

$$\mathbb{E}Z_{\delta\epsilon m}^2 \leq \delta^2 \epsilon^2 m^2 \mathbb{E}\mathbf{1}_{\{\tau < \delta\epsilon m\}} + \mathbb{E}(Z^w)^2,$$

and the exponential decay of $\mathbf{P}(\tau < \delta\epsilon m)$ and Lemma 5.17 concludes the proof. $\square$

LEMMA 5.27. *For any fixed $\delta \in (0, 1/10)$ denote by τ_δ the stopping time*

$$\tau_\delta = \min_{t \geq \delta\epsilon m}\left\{t \text{ is a record minimum of } Y_t\right\} - \delta\epsilon m.$$

Then

$$\mathbb{E}\tau_\delta \leq (2 - \delta)\epsilon m - \frac{1}{4\epsilon^2}.$$

Proof. Define the process $\{X_t\}$ by $X_t = Y_{\delta\epsilon m + t}$ so that

$$\tau_\delta = \min\{t \geq 0 : X_t = -Z_{\delta\epsilon m}\}.$$

Let $\{W_t\}$ be a process defined by $W_0 = X_0$ and with independent increments distributed as $\mathrm{Bin}(m - t - \delta\epsilon m, p) - 1$ and let τ denote the stopping time $\min_t \{W_t = -Z_{\delta\epsilon m}\}$. As usual, X_t can be stochastically bounded above by W_t and hence $\mathbb{E}\tau_\delta \leq \mathbb{E}\tau$ and we are left to estimate $\mathbb{E}\tau$. We have

(5.42) $$\mathbb{E}[W_t - W_{t-1} \mid \mathcal{F}_{t-1}] = (1 - \delta)\epsilon - \frac{t(1 + \epsilon)}{m} - \delta\epsilon^2.$$

Put
$$f(t) = \frac{t^2}{2m} - (1-\delta)\epsilon t - (\delta - \delta^2/2)\epsilon^2 m - \delta\epsilon^2 t + \frac{t(1+\epsilon) + \epsilon t^2}{2m}$$
$$= \frac{[t - (2-\delta)\epsilon m]^2}{2m} + \epsilon[t - (2-\delta)\epsilon m] - \delta\epsilon^2 t + \frac{t(1+\epsilon) + \epsilon t^2}{2m},$$

then by (5.42) we deduce that $M_t = W_t + f(t)$ is a martingale. A direct calculation with (5.4) gives that
$$\mathbb{E}W_0 = \mathbb{E}Y_{\delta\epsilon m} \leq -\delta\epsilon m + \delta\epsilon^2 m - \delta^2\epsilon^2 m/2 + O(\epsilon^3 m),$$

and so we deduce that $\mathbb{E}M_0 \leq C\epsilon^3 m$. Furthermore, we have that $\mathbb{E}\tau = O(\epsilon m)$ since after time $2\epsilon m$ the process becomes subcritical with drift $-\epsilon$. Put $\bar\tau = \tau - (2-\delta)\epsilon m$, then by the above and optional stopping if follows that
$$\frac{\mathbb{E}\bar\tau^2}{2m} + \epsilon\mathbb{E}\bar\tau - \mathbb{E}Z_{\delta\epsilon m} \leq C\epsilon^3 m.$$

This and Lemma 5.26 gives that

(5.43) $$\mathbb{E}\bar\tau \leq \frac{1}{2[(1 - 5\delta - 5\delta\epsilon)\epsilon]} - \frac{\mathbb{E}\bar\tau^2}{2\epsilon m} + O(\epsilon^2 m).$$

Next, we wish to derive a lower bound on $\mathbb{E}\bar\tau^2$. Put $T = \delta m$, then for $t \leq T$ we have that
$$\mathbb{E}\Big[(M_t - M_{t-1})^2\Big] \geq 1 - \delta,$$

hence the process
$$M_{t \wedge T}^2 - (1-\delta)(t \wedge T),$$

is a submartingale and optional stopping gives

(5.44) $$(1-\delta)\mathbb{E}[\tau \wedge T] \leq \mathbb{E}M_{\tau \wedge T}^2.$$

We now bound $\mathbb{E}M_{\tau \wedge T}^2$ from above. We have
$$W_{\tau \wedge T} = -Z_{\delta\epsilon m}\mathbf{1}_{\{\tau \leq T\}} + W_T \mathbf{1}_{\{\tau > T\}}.$$

Thus,
$$\mathbb{E}W_{\tau \wedge T}^2 \leq \mathbb{E}Z_{\delta\epsilon m}^2 + O(m^2)\mathbf{P}(\tau > T).$$

Since after time $\delta m/2$ the process is subcritical with constant negative drift we have that $\mathbf{P}(\tau > T)$ decays exponentially in m. Lemma 5.17 now yields that $\mathbb{E}W_{\tau \wedge T}^2 = \mathbb{E}Z_{\delta\epsilon m}^2 + o(1) = O(\epsilon^{-2})$. Next we estimate $\mathbb{E}f^2(\tau \wedge T)$. Write $\mu = (2-\delta)\epsilon m$ and simplify $f(t)$ to get
$$f(t) = \frac{(t-\mu)^2(1-\epsilon)}{2m} + (t-\mu)\Big[\epsilon - \delta\epsilon^2 + \frac{1+\epsilon}{2m} - \frac{\mu\epsilon}{m}\Big] - \delta\epsilon^2\mu + \frac{1+\epsilon}{2m}\mu - \frac{\epsilon}{2m}\mu^2$$
$$= \frac{(t-\mu)^2(1-\epsilon)}{2m} + (t-\mu)\Big[\epsilon + O(\epsilon^2)\Big] + O(\epsilon^3 m).$$

Hence
$$f^2(t) = \frac{(t-\mu)^4(1-\epsilon)^2}{4m^2} + \frac{(t-\mu)^3(\epsilon + O(\epsilon^2))}{m} + (t-\mu)^2\epsilon^2(1+O(\epsilon)) + (t-\mu)O(\epsilon^4 m).$$

Lemmas 5.23 and 5.25 imply that $\mathbb{E}\bar\tau^k$ is of order $(m/\epsilon)^{k/2}$ and hence the third term on the right hand side is dominant so,

(5.45) $$\mathbb{E}f^2(\tau \wedge T) = (1 + o(1))\epsilon^2 \mathbb{E}\Big[(\tau \wedge T - \mu)^2\Big].$$

We also use Cauchy-Schwartz to estimate
$$\left|\mathbb{E}W_{\tau\wedge T}f(\tau\wedge T)\right| \leq \sqrt{\mathbb{E}W^2_{\tau\wedge T}}\sqrt{\mathbb{E}f^2(\tau\wedge T)} = O(\sqrt{m/\epsilon}) = o(\mathbb{E}f^2(\tau\wedge T)),$$
since $\mathbb{E}W^2_{\tau\wedge T} = O(\epsilon^{-2})$ and $\sqrt{m/\epsilon} = o(\epsilon m)$. We put this and (5.45) into (5.44) and get that
$$(1+o(1))\epsilon^2\mathbb{E}\left[(\tau\wedge T - \mu)^2\right] \geq (1-\delta)\mu - (1-\delta)\mathbb{E}[\tau\wedge T - \mu] = (1+o(1))(1-\delta)\mu,$$
since $\mathbb{E}\bar{\tau} = O(\epsilon^{-1}m)$ and $\mathbf{P}(\tau > T)$ decays exponentially in m. We learn that
$$\mathbb{E}\bar{\tau}^2 \geq (1-o(1))(1-\delta)(2-\delta)\frac{m}{\epsilon}.$$
Putting this into (5.43) gives that if $\delta > 0$ is chosen small enough (but fixed) and m is large enough
$$\mathbb{E}\bar{\tau} \leq -\frac{1}{4\epsilon^2},$$
concluding the proof of the lemma. □

Proof of Theorem 5.15. Part (i) follows immediately from Lemma 5.27 since $|\mathcal{C}_{\delta\epsilon m}| \leq \delta\epsilon m + \tau_\delta$. To prove (ii) we proceed as in the proof of Lemma 5.10 and write
$$\mathbb{E}\sum_{\mathcal{C}_j \neq \mathcal{C}_{\delta\epsilon m}}|\mathcal{C}_j|^k = \mathbb{E}\sum_{\mathcal{C}_j \neq \mathcal{C}_{\delta\epsilon m}}|\mathcal{C}_j|^k\mathbf{1}_{\{\mathcal{C}_{\delta\epsilon m}\leq 1.5\epsilon m\}} + \mathbb{E}\sum_{j\geq 2}|\mathcal{C}_j|^k\mathbf{1}_{\{\mathcal{C}_{\delta\epsilon m}\geq 1.5\epsilon m\}}.$$
Lemma 5.23 shows that $\mathbf{P}(\mathcal{C}_{\delta\epsilon m} \leq 1.5\epsilon m) \leq Ce^{-c\epsilon^3 m}$ and so FKG inequality gives
$$\mathbb{E}\sum_{\mathcal{C}_j \neq \mathcal{C}_{\delta\epsilon m}}|\mathcal{C}_j|^k\mathbf{1}_{\{\mathcal{C}_{\delta\epsilon m}\leq 1.5\epsilon m\}} \leq Ce^{-c\epsilon^3 m}\mathbb{E}\sum_j|\mathcal{C}_j|^k = O(m\epsilon^{-2k+3}),$$
by part (i) of Lemma 5.13. The second term is handled as in the proof of Lemma 5.10 by conditioning on $\mathcal{C}_{\delta\epsilon m}$ and using Lemma 5.13 for the remaining subcritical graph. □

CHAPTER 6

Supercritical case

In this section we show that the mixing time of the Swendsen-Wang chain is $\Theta(\log n)$ in the supercritical case $c > 2$. This is part (i) of Theorem 2.1. Let $\{X_t\}_{t \geq 0}$ be the one dimensional chain defined in (4.1) and write $x_0 = X_0/n$. For $x > \frac{2}{c} - 1$ (so that $\frac{c(1+x)}{2} > 1$), define

$$\phi(x) = \beta\left(\frac{c(1+x)}{2}\right)\frac{1+x}{2}, \tag{6.1}$$

where $\beta(\cdot)$ is defined in (5.9). Since $\beta : \mathbb{R}^+ \to \mathbb{R}$ we have that $\phi : [-1, \infty] \to \mathbb{R}$. We begin with some preparations for the proof.

LEMMA 6.1. *For $c > 2$, there exists an unique fixed point $\gamma_0 \in (1 - \frac{2}{c}, 1)$ of $\phi(x)$. Furthermore, we have*

$$\frac{1}{2} < \phi'(x) < 1 \qquad \text{for } x > 2c^{-1} - 1, \tag{6.2}$$

and there exists a constant $\delta \in (0, 1)$ such that for every $x \in [0, 1] \setminus \{\gamma_0\}$ we have

$$\frac{1}{2} < \frac{\phi(x) - \gamma_0}{x - \gamma_0} \leq \delta. \tag{6.3}$$

THEOREM 6.2. *There exist constants $\delta \in (0, 1)$ and $B > 0$ such that*

$$\mathbb{E}(X_1 - \gamma_0 n)^2 \leq \delta(X_0 - \gamma_0 n)^2 + Bn. \tag{6.4}$$

PROPOSITION 6.3. *We have*

$$\mathbb{E}\left(X_1 - \phi(x_0)n \mid X_0 \in [\gamma_0 n, n]\right)^2 = O(n).$$

PROPOSITION 6.4. *If X is distributed as the stationary distribution of the magnetization Swendsen-Wang chain, then*

$$\mathbb{E}(X - \gamma_0 n)^2 = O(n).$$

THEOREM 6.5. *Suppose X_0, Y_0 are two magnetization Swendsen-Wang chains such that $X_0, Y_0 \in [\gamma_0 n - A\sqrt{n}, \gamma_0 n + A\sqrt{n}]$ where A is a constant, we can couple X_1 and Y_1 such that $X_1 = Y_1$ with probability $\Omega(1)$ (which may depend on A).*

Proof of part (i) of Theorem 2.1: Rearranging Theorem 6.2 and taking expectations gives

$$\mathbb{E}(X_{t+1} - \gamma_0 n)^2 - \frac{B}{1-\delta}n \leq \delta\left[\mathbb{E}(X_t - \gamma_0 n)^2 - \frac{B}{1-\delta}n\right]$$

for all t. We apply this inductively and get

$$\mathbb{E}(X_{C\log n} - \gamma_0 n)^2 - \frac{B}{1-\delta}n \leq \delta^{C\log n}\left[\mathbb{E}(X_0 - \gamma_0 n)^2 - \frac{B}{1-\delta}n\right].$$

Hence, when $C = C(\delta)$ is large enough we get that
$$\mathbb{E}(X_{C\log n} - \gamma_0 n)^2 = O(n),$$
and so Markov's inequality gives
$$(6.5) \quad \mathbf{P}(|X_{C\log n} - \gamma_0 n| \leq A\sqrt{n}) \geq \frac{3}{4}$$
for some large constant A. Let X'_t be a magnetization SW chain starting at stationarity. By Theorem 6.4 and Markov's inequality we have
$$(6.6) \quad \mathbf{P}(|X'_{C\log n} - \gamma_0 n| \leq A\sqrt{n}) \geq \frac{3}{4}$$
for some large constant A. Now, to couple X_t and X'_t we first run them independently until time $C \log n$. By (6.5) and (6.6), we have that $X_{C\log n}, X'_{C\log n} \in [\gamma_0 n - A\sqrt{n}, \gamma_0 n + A\sqrt{n}]$ with probability at least $1/2$. By Theorem 6.5, we can couple $X_{C\log n+1}$ and $X'_{C\log n+1}$ such that $X_{C\log n+1} = X'_{C\log n+1}$ with probability $\Omega(1)$. Then by Lemma 4.1, we have that $\{\sigma_t\}$ and $\{\sigma'_t\}$ can be coupled such that $\sigma_t = \sigma'_t$ in $O(\log n)$ steps with probability $\Omega(1)$. The upper bound of mixing time follows from Lemma 3.2.

For the lower bound, we will show that if $X_0 = n$, then
$$\|X_{\alpha \log n} - X_\pi\|_{TV} \geq 1/4$$
for some small constant $\alpha > 0$, where X_π is the stationary distribution of magnetization Swendsen-Wang chain. By (6.3), we have that
$$\mathbf{P}\left(X_{t+1} - \gamma_0 n \leq \frac{1}{4}(X_t - \gamma_0 n)\right) \leq \mathbf{P}\left(X_{t+1} - \gamma_0 n \leq \frac{1}{2}(\phi(X_t/n)n - \gamma_0 n)\right)$$
$$= \mathbf{P}\left(X_{t+1} - \phi(X_t/n)n \leq \frac{1}{2}(\gamma_0 n - \phi(X_t/n)n)\right).$$
When $X_t \geq \gamma_0 n$ we have that $\phi(X_t/n)n \geq \gamma_0 n$ by (6.3), hence Proposition 6.3 and Markov's inequality imply that
$$(6.7) \quad \mathbf{P}\left(X_{t+1} - \gamma_0 n \leq \frac{1}{4}(X_t - \gamma_0 n) \mid X_t \geq \gamma_0 n\right) \leq \frac{O(n)}{(\phi(X_t/n)n - \gamma_0 n)^2}.$$
Furthermore, if $X_t - \gamma_0 n \geq n^{\frac{3}{4}}$, then $\phi(X_t/n)n - \gamma_0 n \geq n^{\frac{3}{4}}/2$ by (6.3). Plugging this into (6.7) gives
$$(6.8) \quad \mathbf{P}\left(X_{t+1} - \gamma_0 n \geq \frac{1}{4}(X_t - \gamma_0 n) \big| X_t - \gamma_0 n \geq n^{\frac{3}{4}}\right) \geq 1 - O(n^{-\frac{1}{2}}).$$
Starting from $X_0 = n$, by applying (6.8) iteratively we have
$$(6.9) \quad \mathbf{P}\left(X_{\alpha \log n} - \gamma_0 n \geq n^{\frac{3}{4}}\right) \geq (1 - O(n^{-\frac{1}{2}}))^{\alpha \log n} = 1 - o(1),$$
when $\alpha > 0$ is small enough constant. On the other hand, by Proposition 6.4 and the Markov's inequality, we have $\mathbf{P}\left(|X_\pi - \gamma_0 n| \geq A\sqrt{n}\right) \leq \frac{1}{4}$ for some constant A. Putting the two inequalities together, we get
$$(6.10) \quad \|X_{\alpha \log_4 n} - X_\pi\|_{TV} \geq \frac{3}{4} - o(1) \geq \frac{1}{4},$$
which gives a lower bound on the mixing time of magnetization SW chain X_t. This concludes the proof since any lower bound of the mixing time of X_t implies the same lower bound of mixing time of σ_t. $\square$

Proof of Lemma 6.1: By the definition of $\beta(\cdot)$ in equation (5.9), we know $\phi(x)$ is the positive solution of

$$1 - e^{-c\phi(x)} = \frac{2\phi(x)}{x+1} \tag{6.11}$$

for all $x > \frac{2}{c} - 1$. Taking derivative of both sides yields

$$ce^{-c\phi}\phi' = \frac{2(x+1)\phi' - 2\phi}{(x+1)^2}.$$

By plugging in $x + 1 = \frac{2\phi}{1-e^{-c\phi}}$ we get

$$\phi' = \frac{1 - 2e^{-c\phi} + e^{-2c\phi}}{2(1 - e^{-c\phi} - c\phi e^{-c\phi})}. \tag{6.12}$$

By (6.12), we have that $\frac{1}{2} < \phi'$ if and only if $e^{-c\phi} > 1 - c\phi$ which is true for all $c\phi > 0$. We also have that

$$\phi' < 1 \iff c\phi < \sinh(c\phi) \tag{6.13}$$

which holds for all $c\phi > 0$.

Since $c > 2$ (which implies $\frac{2}{c} - 1 < 0$), we have that $\phi' < 1$ for all $x \in [0,1]$. Since ϕ' is continuous, we have a constant $\delta_1 \in (0,1)$ such that $\phi'(x) < \delta_1$ for all $x \in [0,1]$. Note that $\phi(0) = \frac{1}{2}\beta(\frac{c}{2}) > 0$, $\phi(1) = \beta(c) < 1$ and ϕ is strictly increasing in $[0,1]$, we have by Rolle's theorem that there exists an unique point $\gamma_0 \in (0,1)$ such that $\phi(\gamma_0) = \gamma_0$. By plugging in $x = 1 - 2/c$ into (6.1) and the definition of $\beta(\cdot)$, we get

$$\phi(1 - 2/c) > 1 - 2/c \iff \beta(c-1) > 1 - \frac{1}{c-1}$$
$$\iff e^{-(c-2)} < \frac{1}{c-1}$$

which is always true for $c > 2$. It follows immediately that $\gamma_0 > 1 - \frac{2}{c}$. $\square$

Recall that given $X_0 = x_0 n$, we have that X_1 is distributed as in (4.2). To prove Theorem 6.2 we first state a useful lemma.

LEMMA 6.6. *Let $c > 2$.*

(i) *There exists a non-negative function $h(\cdot)$ with $h(\epsilon) \to 0$ as $\epsilon \to 0$ such that if $|x_0 - (1 - \frac{2}{c})| \le \epsilon$, then $\mathbb{E}\left(\sum_{j\ge 1} |\mathcal{C}_j^-|^2\right) \le h(\epsilon) n^2$.*

(ii) *For any fixed $\epsilon > 0$, if $x_0 \in [0, (1 - \frac{2}{c} - \epsilon)]$, then $\mathbb{E}\left(\sum_{j\ge 1} |\mathcal{C}_j^-|^2\right) \le (\phi^2(-x_0) + o(1))n^2$.*

(iii) *For any fixed $\epsilon > 0$, if $x_0 \in [1 - \frac{2}{c} + \epsilon, 1]$, then $\mathbb{E}\left(\sum_{j\ge 1} |\mathcal{C}_j^-|^2\right) \le O(n)$.*

Proof of Theorem 6.2: By (4.2), we have

$$\mathbb{E}X_1^2 = \mathbb{E}\left(\sum_{j\ge 1} |\mathcal{C}_j^+|^2\right) + \mathbb{E}\left(\sum_{j\ge 1} |\mathcal{C}_j^-|^2\right) \tag{6.14}$$

and
$$\begin{aligned}
\mathbb{E}X_1 &= \mathbb{E}\Big|\sum_{j\geq 1}\epsilon_j|\mathcal{C}_j^+| + \sum_{j\geq 1}\epsilon_j'|\mathcal{C}_j^-|\Big| = \mathbb{E}\Big|\epsilon_1\big(\sum_{j\geq 1}\epsilon_j|\mathcal{C}_j^+| + \sum_{j\geq 1}\epsilon_j'|\mathcal{C}_j^-|\big)\Big| \\
&\geq \mathbb{E}\Big(\epsilon_1\big(\sum_{j\geq 1}\epsilon_j|\mathcal{C}_j^+| + \sum_{j\geq 1}\epsilon_j'|\mathcal{C}_j^-|\big)\Big) \\
(6.15)\quad &= \mathbb{E}\Big[|\mathcal{C}_1^+| + \epsilon_1\big(\sum_{j\geq 2}\epsilon_j|\mathcal{C}_j^+| + \sum_{j\geq 1}\epsilon_j'|\mathcal{C}_j^-|\big)\Big] = \mathbb{E}|\mathcal{C}_1^+|.
\end{aligned}$$

Combining (6.14) and (6.15), we get

$$(6.16)\quad \mathbb{E}\Big(X_1 - \gamma_0 n\Big)^2 \leq \mathbb{E}\Big(\sum_{j\geq 1}|\mathcal{C}_j^+|^2\Big) + \mathbb{E}\Big(\sum_{j\geq 1}|\mathcal{C}_j^-|^2\Big) - 2\gamma_0 n \cdot \mathbb{E}|\mathcal{C}_1^+| + \gamma_0^2 n^2.$$

The random graph $G(\frac{1+x_0}{2}n, \frac{c}{n})$ is supercritical with $\theta = \frac{1+x_0}{2}n\frac{c}{n} \geq \frac{c}{2} > 1$. By Corollary 5.6 we have

$$(6.17)\quad \mathbb{E}\Big(\sum_{j\geq 1}|\mathcal{C}_j^+|^2\Big) \leq \Big(\mathbb{E}|\mathcal{C}_1^+|\Big)^2 + O(n).$$

Plugging (6.17) into (6.16), we get

$$(6.18)\quad \mathbb{E}\Big(X_1 - \gamma_0 n\Big)^2 \leq \Big(\mathbb{E}|\mathcal{C}_1^+| - \gamma_0 n\Big)^2 + \mathbb{E}\Big(\sum_{j\geq 1}|\mathcal{C}_j^-|^2\Big) + O(n).$$

By Corollary 5.5, we have $\big|\mathbb{E}|\mathcal{C}_1^+| - \phi(x_0)n\big| \leq O(\sqrt{n})$. Thus,

$$\begin{aligned}
\Big(\mathbb{E}|\mathcal{C}_1^+| - \gamma_0 n\Big)^2 &\leq \big|\mathbb{E}|\mathcal{C}_1^+| - \phi(x_0)n\big|^2 + \big|\phi(x_0)n - \gamma_0 n\big|^2 \\
&\quad + 2\big|\mathbb{E}|\mathcal{C}_1^+| - \phi(x_0)n\big|\big|\phi(x_0)n - \gamma_0 n\big| \\
(6.19)\quad &\leq \big|\phi(x_0)n - \gamma_0 n\big|^2 + O(\sqrt{n})\big|\phi(x_0)n - \gamma_0 n\big| + O(n).
\end{aligned}$$

Applying Lemma 6.1 gives that

$$(6.20)\quad \Big(\mathbb{E}|\mathcal{C}_1^+| - \gamma_0 n\Big)^2 \leq \delta_1^2|x_0 - \gamma_0|^2 n^2 + |x_0 - \gamma_0|O(n^{3/2}) + O(n).$$

If $|x_0 - \gamma_0| = O(n^{-\frac{1}{2}})$, then $|x_0 - \gamma_0|n^{3/2} = O(n)$. If $|x_0 - \gamma_0|n^{\frac{1}{2}} \to \infty$, we have $|x_0 - \gamma_0|O(n^{3/2}) = o(|x_0 - \gamma_0|^2 n^2)$. Plugging these back into (6.20), we get

$$(6.21)\quad \mathbb{E}(X_1 - \gamma_0 n)^2 \leq (\delta_1^2 + o(1))|x_0 - \gamma_0|^2 n^2 + O(n) + \mathbb{E}\Big(\sum_{j\geq 1}|\mathcal{C}_j^-|^2\Big).$$

To estimate $\mathbb{E}\big(\sum_{j\geq 1}|\mathcal{C}_j^-|^2\big)$, choose a small constant ϵ such that $\delta_1^2 + h(\epsilon) < 1$ where $h(\cdot)$ is defined in part (i) of Lemma 6.6. If $\big|x_0 - (1 - \frac{2}{c})\big| < \epsilon$, we have that

$$(6.22)\quad \mathbb{E}(X_1 - \gamma_0 n)^2 \leq (\delta_1^2 + h(\epsilon))|x_0 - \gamma_0|^2 n^2 + O(n)$$

by plugging part (i) of Lemma 6.6 into (6.21).

If $x_0 \in [0, (1 - \frac{2}{c} - \epsilon)]$, we have that $|\phi(x_0) - \gamma_0|$ is uniformly bounded from below by Lemma 6.1. As a result, we have that $\Big(\mathbb{E}|\mathcal{C}_1^+| - \gamma_0 n\Big)^2 \leq (\phi(x_0) - \gamma_0)^2 n^2 + O(n)$

in (6.19). Plugging this and part (ii) of Lemma 6.6 into (6.19) gives

(6.23) $\quad \mathbb{E}(X_1 - \gamma_0 n)^2 \leq ((\phi(x_0) - \gamma_0)^2 + \phi^2(-x_0) + o(1))n^2 + O(n).$

By Lemma 6.1 and Rolle's Theorem, we have

$$\frac{\phi(x_0) - \phi(-x_0)}{2x_0} \geq \frac{1}{2},$$

which leads to $\phi^2(-x_0) \leq \left(\phi(x_0) - x_0\right)^2$. This gives

(6.24) $\quad \dfrac{\left(\phi(x_0) - \gamma_0\right)^2 + \phi^2(-x_0)}{(x_0 - \gamma_0)^2} \leq \dfrac{\left(\phi(x_0) - \gamma_0\right)^2 + \left(\phi(x_0) - x_0\right)^2}{(x_0 - \gamma_0)^2} < 1,$

since $x_0 < \phi(x_0) < \gamma_0$. The left hand side of (6.24) is smaller than 1 for all $x_0 \in [0, (1 - \frac{2}{c} - \beta)]$, so it is smaller than some constant $\delta_2 < 1$ uniformly. Plugging this into (6.23), we get

(6.25) $\quad \mathbb{E}(X_1 - \gamma_0 n)^2 \leq \delta_2 (x_0 - \gamma_0)^2 n^2 + O(n).$

If $x_0 \in [1 - \frac{2}{c} + \epsilon, 1]$, we plug (iii) of Lemma 6.6 into (6.21) and obtain

(6.26) $\quad \mathbb{E}(X_1 - \gamma_0 n)^2 \leq (\delta_1^2 + o(1))|x_0 - \gamma_0|^2 n^2 + O(n).$

Combining (6.22),(6.25) and (6.26) concludes our proof. $\square$

Proof of Lemma 6.6. We begin with case (ii). In this regime, the random graph $G(\frac{1-x_0}{2}n, \frac{c}{n})$ is supercritical with $\theta > 1 + \frac{c\epsilon}{2}$. In the same way we obtained (6.17) we also have

(6.27) $\quad \mathbb{E}\Big(\sum_{j\geq 1}|\mathcal{C}_j^-|^2\Big) \leq \Big(\mathbb{E}|\mathcal{C}_1^-|\Big)^2 + O(n).$

By Corollary 5.5 we have that $|\mathbb{E}|\mathcal{C}_1^-| - \phi(-x_0)n| \leq O(\sqrt{n})$ showing that

$$\mathbb{E}\Big(\sum_{j\geq 1}|\mathcal{C}_j^-|^2\Big) \leq \Big(\phi(-x_0)n + O(\sqrt{n})\Big)^2 + O(n)$$

(6.28) $\quad \begin{aligned} &\leq \phi^2(-x_0)n^2 + O(n) + \phi(-x_0)O(n^{3/2}) \\ &= (\phi^2(-x_0) + o(1))n^2, \end{aligned}$

since $|\phi(x_0)|$ is uniformly bounded from below, as required.

We now prove case (i). Note that we have

$$\mathbb{E}\Big(\sum_{j\geq 1}|\mathcal{C}_j^-|^2\Big) = \Big(\frac{1-x_0}{2}n\Big)\mathbb{E}|\mathcal{C}_v|$$

as in (5.8). Since $\mathbb{E}|\mathcal{C}_v|$ is decreasing in x_0, we have that $\mathbb{E}\Big(\sum_{j\geq 1}|\mathcal{C}_j^-|^2\Big)$ reaches its maximum at $x_0 = 1 - \frac{2}{c} - \epsilon$. Plugging in this value into (6.28) gives

(6.29) $\quad \mathbb{E}\Big(\sum_{j\geq 1}|\mathcal{C}_j^-|^2\Big) \leq \Big(\beta^2(1 + \frac{\epsilon c}{2}) + o(1)\Big)n^2$

Note that $\beta(x) \to 0$ as $x \to 1$, so we can take $h(x) = \beta^2(1 + \frac{cx}{2}) + o(1)$.

To prove case (iii) note that $\frac{c(1-x_0)}{2} \leq 1 - \frac{\epsilon c}{2}$, so the random graph $G(\frac{1-x_0}{2}n, \frac{c}{n})$ is subcritical in this regime with θ bounded from above away from 1. Applying Lemma 5.3, we get

$$\mathbb{E}\Big[\sum_{j\geq 1} |\mathcal{C}_j^-|^2\Big] = O(n). \tag{6.30}$$

$\square$

Proof of Proposition 6.3: Note that (6.18) is valid for all $\gamma_0 \in [0,1]$ and in particular for $\phi(x_0)$. Thus,

$$\mathbb{E}\Big(X_1 - \phi(x_0)n\Big)^2 \leq \Big(\mathbb{E}|\mathcal{C}_1^+| - \phi(x_0)n\Big)^2 + \mathbb{E}\Big(\sum_{j\geq 1}|\mathcal{C}_j^-|^2\Big) + O(n). \tag{6.31}$$

Recall that $x_0 \geq \gamma_0 > 1 - \frac{2}{c}$. By Corollary 5.5 we have that $\big(\mathbb{E}|\mathcal{C}_1^+| - \phi(x_0)n\big)^2 = O(n)$. The random graph of $G(\frac{1-x_0}{2}n, \frac{c}{n})$ is in regime (iii) of Lemma 6.6. Plugging (6.30) into (6.31), we get

$$\mathbb{E}\Big(X_1 - \phi(x_0)n\Big)^2 = O(n), \tag{6.32}$$

as required. $\square$

Proof of Proposition 6.4: If X_0 follows the stationary distribution of the magnetization SW chain, so does X_1. Taking expectation of both sides of (6.4) gives

$$\mathbb{E}(X_1 - \gamma_0 n)^2 \leq \delta \mathbb{E}(X_0 - \gamma_0 n)^2 + Bn,$$

as required. $\square$

To prove Theorem 6.5 we need the following lemma.

LEMMA 6.7. *Let Y and Z be two random variables distributed as the sum of n independent random $\pm$ signs. Then for any fixed constant a, there exists a constant $\kappa(a) \in (0,1]$ such that for any $-a\sqrt{n} \leq y \leq a\sqrt{n}$, we can couple Y and Z such that $Y - y = Z$ with probability at least κ.*

Proof. Direct corollary of the local central limit theorem of simple random walk.
$\square$

Proof of Theorem 6.5. To couple X_1 and Y_1, we first apply the percolation step of the Swendsen-Wang dynamics in both chains independently. By Lemma 5.7, with probability $1 - O(\frac{1}{n})$, the number of isolated points after percolation is bigger than $\frac{1}{3e^c}n$ in both chains. Conditioned on this, we assign each component a $\pm$ spin using the following procedure.

First assign the spins of components independently in descending order of their size until there are $\frac{1}{3e^c}n$ components left. Note the remaining components are all isolated vertices. Denote by $\bar{X}_1$ and $\bar{Y}_1$ as the absolute value of the sum of spins at this time respectively.

Note that (6.15) and (6.16) are still valid if we replace X_1 by $\bar{X}_1$. Consequently, Theorem 6.2 is also valid if replacing X_1 by $\bar{X}_1$. Hence, since $|X_0 - \gamma_0 n| \leq A\sqrt{n}$ we have

$$\mathbb{E}(\bar{X}_1 - \gamma_0 n)^2 = O(n).$$

By Markov's inequality, there exists a constant A_1 such that

$$\mathbf{P}\left(|\bar{X}_1 - \gamma_0 n| \geq A_1 \sqrt{n}\right) \leq \frac{1}{4}, \tag{6.33}$$

and similarly

$$\mathbf{P}\left(|\bar{Y}_1 - \gamma_0 n| \geq A_1 \sqrt{n}\right) \leq \frac{1}{4}. \tag{6.34}$$

Consider the event
$$\mathcal{A} := \{|\bar{X}_1 - \gamma_0 n| < A_1 \sqrt{n}\} \cap \{|\bar{Y}_1 - \gamma_0 n| < A_1 \sqrt{n}\} \cap \{\text{There are at least } \tfrac{n}{3e^c} \text{ isolated vertices}\}.$$
By (6.33) and (6.34) we have that $\mathbf{P}(\mathcal{A}) \geq \frac{1}{4}$.

Conditioned on $\mathcal{A}$, we have $|\bar{X}_1 - \bar{Y}_1| \leq 2A_1 \sqrt{n}$. Denote by $\hat{X}_1$ and $\hat{Y}_1$ the sum of spins of the rest of the components (all of them being isolated vertices) of the two chains respectively. Note $\hat{X}_1$ and $\hat{Y}_1$ are i.i.d. sums of $\pm$ spins. By Lemma 6.7 we can couple $\hat{X}_1$ and $\hat{Y}_1$ so that $\hat{X}_1 + \bar{X}_1 = \hat{Y}_1 + \bar{Y}_1$ with probability $\Omega(1)$. Finally, notice that $X_1 \stackrel{(d)}{=} |\bar{X}_1 + \hat{X}_1|$ and $Y_1 \stackrel{(d)}{=} |\bar{Y}_1 + \hat{Y}_1|$, concluding the proof. $\square$

CHAPTER 7

Subcritical case

In this section, we prove that in the subcritical case $c < 2$, the mixing time of Swendsen-Wang chain is $\Theta(1)$. This is part (iii) of Theorem 2.1.

LEMMA 7.1. *For $c \in (1, 2)$ there exists a constant $\delta \in (0, 1)$ such that for all $x \in [\frac{2}{c} - 1, 1]$, we have*

(7.1) $$\frac{\phi(x)}{x} \leq \delta$$

where $\phi(\cdot)$ is defined in (6.1).

THEOREM 7.2. *There exist two constants $\delta \in (0, 1)$ and $B > 0$ such that*

(7.2) $$\mathbb{E}(X_1^2 \mid X_0) \leq \delta X_0^2 + Bn.$$

Moreover, if $0 \leq x_0 \leq \frac{1}{c} - \frac{1}{2}$, we have

(7.3) $$\mathbb{E} X_1^2 \leq Bn.$$

To get the constant upper bound of mixing time we need to consider the following two-dimensional chain. Let G_1 be a fixed subset of the vertices and G_2 its complement. Let (Y_t, Z_t) be a two-dimensional Markov chain, where Y_t record the number of vertices with positive spin in G_1 and Z_t record the number vertices with positive spin in G_2.

PROPOSITION 7.3. *Let (Y_t, Z_t) and $(\widetilde{Y}_t, \widetilde{Z}_t)$ be two two-dimensional chains as defined above. Suppose $Y_0 + Z_0$ and $\widetilde{Y}_0 + \widetilde{Z}_0$ lie in the window $I = [\frac{n}{2} - A\sqrt{n}, \frac{n}{2} + A\sqrt{n}]$ where A is a constant. Then we can couple (Y_1, Z_1) and $(\widetilde{Y}_1, \widetilde{Z}_1)$ such that $(Y_1, Z_1) = (\widetilde{Y}_1, \widetilde{Z}_1)$ with probability $\Omega(1)$ (which may depend on A).*

Proof of part (iii) of Theorem 2.1: For any starting configuration σ, let G_1 be the vertices with positive spin and G_2 be its complement. Let X_t be the magnetization chain and (Y_t, Z_t) be the two-dimensional chain as described above. As usual $\mathbf{P}$ and π are the transition matrix and the stationary distribution of the Swendsen-Wang chain, respectively, and let $\widetilde{\mathbf{P}}$ and $\widetilde{\pi}$ be the corresponding transition matrix and stationary distribution of (Y_t, Z_t), respectively. By symmetry, configurations with same two-dimensional chain value have same distributions for any t. Consequently

(7.4) $$\|\sigma \mathbf{P}^t - \pi\|_{TV} = \|(|G_1|, 0)\widetilde{\mathbf{P}}^t, \widetilde{\pi}\|_{TV}.$$

Thus, by Lemma 3.2 it suffices to couple the chains (Y_t, Z_t) and $(\widetilde{Y}_t, \widetilde{Z}_t)$ such that they meet with probability $\Omega(1)$ in time $t = \Theta(1)$. By Lemma 7.2, we have

$$\mathbb{E}(X_{t+1}^2) - \frac{B}{1-\delta}n \leq \delta\Big[\mathbb{E}(X_t^2) - \frac{B}{1-\delta}n\Big].$$

Applying this inductively we get

$$\mathbb{E}(X_t^2) - \frac{B}{1-\delta}n \leq \delta^t \mathbb{E}(X_0^2) \leq \delta^t n^2.$$

For $t \geq 2\log_\delta \frac{1}{8}(\frac{1}{c} - \frac{1}{2})$ and large n, we have

$$\mathbb{E}(X_t^2) \leq \frac{1}{4}\left(\frac{1}{c} - \frac{1}{2}\right)^2 n^2.$$

For such t Markov's inequality gives

(7.5) $$\mathbf{P}\left(X_t \geq \left(\frac{1}{c} - \frac{1}{2}\right)n\right) \leq \frac{1}{4}.$$

By Theorem 7.2 and Markov's inequality, if $X_t \in [0, (\frac{1}{c} - \frac{1}{2})n]$, then $X_{t+1} \in [0, A\sqrt{n}]$ with probability at least $1/2$ for some large constant A. Combining this and (7.5), we have that after constant number of steps, the chain X_t will jump into the window $I = [0, A\sqrt{n}]$ with probability $\Omega(1)$.

For any two Swendsen-Wang chains σ and $\tilde{\sigma}$, Let X_t and $\widetilde{X}_t$ be the corresponding magnetization chains. Running the two Swenden-Wang dynamics independently first, by the argument above, we have that X_t and $\widetilde{X}_t$ both jump into $[0, A\sqrt{n}]$ after constant steps with probability $\Omega(1)$. By Proposition 7.3, we can couple the two two-dimensional chains so that $(Y_t, Z_t) = (\widetilde{Y}_t, \widetilde{Z}_t)$ with probability $\Omega(1)$, which concludes the whole proof. □

Proof of Lemma 7.1: Note ϕ is differentiable on $[\frac{2}{c} - 1, 1]$. Recalling (6.13), we have $\phi' < 1$ for all $x > \frac{2}{c} - 1$. By Rolle's Theorem, we have $\phi(x) - 0 \leq x - (\frac{2}{c} - 1)$ for all $x > \frac{2}{c} - 1$. So

$$\frac{\phi(x)}{x} \leq 1 - \frac{\frac{2}{c} - 1}{x} \leq 1 - \left(\frac{2}{c} - 1\right)$$

for all $x \in [\frac{c}{2} - 1, 1]$. □

Proof of Theorem 7.2: We use the fact that (6.14) is still valid. The random graph $G(\frac{1-x_0}{2}n, \frac{c}{n})$ is subcritical with $\theta = (\frac{1-x_0}{2}n)\frac{c}{n} = \frac{c}{2}$. By Lemma 5.3, we have

(7.6) $$\mathbb{E}\left(\sum_{j \geq 1} |\mathcal{C}_j^-|^2\right) = O(n).$$

If $c < 1$, the random graph $G(\frac{1+x_0}{2}n, \frac{c}{n})$ is subcritical with $\theta = (\frac{1+x_0}{2}n)\frac{c}{n} \leq c < 1$. By Lemma 5.3, we have

(7.7) $$\mathbb{E}\left(\sum_{j \geq 1} |\mathcal{C}_j^+|^2\right) = O(n).$$

If $c \geq 1$, then let $\epsilon > 0$ be a small constant that we will determine later and consider the following three cases.

(i) $0 \leq x_0 \leq \frac{2}{c} - 1 - \epsilon$. In this case, the random graph $G(\frac{1+x_0}{2}n, \frac{c}{n})$ is subcritical with $\theta \leq 1 - \frac{\epsilon c}{2}$. By Lemma 5.3,

(7.8) $$\mathbb{E}\left(\sum_{j \geq 1} |\mathcal{C}_j^+|^2\right) = O(n).$$

(ii) $\frac{2}{c} - 1 + \epsilon \leq x_0 \leq 1$ (in case $c > 1$). In this case, the random graph $G(\frac{1+x_0}{2}n, \frac{c}{n})$ is supercritical with $\theta \geq 1 + \frac{\epsilon c}{2}$. By Corollary 5.6, we have

$$\mathbb{E}\Big(\sum_{j\geq 1} |\mathcal{C}_j^+|^2\Big) \leq (\mathbb{E}|\mathcal{C}_1^+|)^2 + O(n)$$
$$= \Big(\phi(x_0)n\Big)^2 + \Big(\mathbb{E}|\mathcal{C}_1^+| - \phi(x_0)n\Big)\Big(\mathbb{E}|\mathcal{C}_1^+| + \phi(x_0)n\Big) + O(n).$$

By Corollary 5.5, we have $\big|\mathbb{E}|\mathcal{C}_1^+| - \phi(x_0)n\big| = O(\sqrt{n})$. By Lemma 7.1, we have $\phi(x_0)n \leq \delta x_0 n$. So we have

$$\mathbb{E}\Big(\sum_{j\geq 1} |\mathcal{C}_j^+|^2\Big) \leq \delta^2 x_0^2 n^2 + O(n^{3/2}) \leq (\delta^2 + o(1))x_0^2 n^2.$$

(iii) $\frac{2}{c} - 1 - \epsilon \leq x_0 \leq \frac{2}{c} - 1 + \epsilon$ (or $1 - \epsilon \leq x_0 \leq 1$ in case $c = 1$). Recall that $\mathbb{E}\Big(\sum_{j\geq 1} |\mathcal{C}_j^+|^2\Big) = \frac{1+x_0}{2}n\mathbb{E}|C_v|$. So $\mathbb{E}\Big(\sum_{j\geq 1} |\mathcal{C}_j^+|^2\Big)$ reaches its maximum at $x_0 = \frac{2}{c} - 1 + \epsilon$ for $1 < c < 2$ or $x_0 = 1$ for $c = 1$. In the former case, by the estimate in case (ii), we get

$$\mathbb{E}\Big(\sum_{j\geq 1} |\mathcal{C}_j^+|^2\Big) \leq (\delta^2 + o(1))(\frac{2}{c} - 1 + \epsilon)^2 n^2.$$

Now we choose ϵ to be small enough such that $\delta^2 \Big(\frac{2/c-1+\epsilon}{2/c-1-\epsilon}\Big)^2 < 1$, then we choose a constant δ_1 such that $\delta^2 \Big(\frac{2/c-1+\epsilon}{2/c-1-\epsilon}\Big)^2 < \delta_1 < 1$. Then we have

$$\mathbb{E}\Big(\sum_{j\geq 1} |\mathcal{C}_j^+|^2\Big) \leq \delta x_0^2 n^2.$$

In the latter case, by Theorem 1 of [25], we have that

$$\mathbb{E}\Big(\sum_{j\geq 1} |\mathcal{C}_j^+|^2\Big) = o(n^2).$$

The Lemma follows from combining case (i), (ii) and (iii). $\square$

Proof of Proposition 7.3: Suppose without lost of generality that $|G_2| \leq |G_1|$. Since $Y_0 + Z_0 \in I$, the random graphs $G(Y_0+Z_0, \frac{c}{n})$ and $G(n-(Y_0+Z_0), \frac{c}{n})$ are both subcritical for large n. The same is true for the chain $(\widetilde{Y}_t, \widetilde{Z}_t)$. In the first chain after the percolation step, denote by $\{\mathcal{A}_j\}_{j\geq 1}$ and $\{\mathcal{B}_j\}_{j\geq 1}$ the components with vertices completely in G_1 and G_2 respectively. Note that there are also components that have vertices in both G_1 and G_2. Denote such components by $\{\mathcal{C}_j\}_{j\geq 1}$. In the second chain, we denote by $\widetilde{\mathcal{A}_j}, \widetilde{\mathcal{B}_j}$ and $\widetilde{\mathcal{C}_j}$ to be these components. Lemma 5.7 implies that for some $c > 0$ with probability $\Omega(1)$ we have that the number of isolated vertices in $\{A_j\}$ is at least $c|G_1|$ and at least $c|G_2|$ for $\{\widetilde{A}_j\}$. Denote this event by $\mathcal{A}$.

Furthermore, by Lemma 5.3 we have

$$\mathbb{E}(\sum_{j\geq 1}|\mathcal{A}_j|^2 + \sum_{j\geq 1}|\mathcal{C}_j \cap G_1|^2) = O(|G_1|), \tag{7.9}$$

$$\mathbb{E}(\sum_{j\geq 1}|\mathcal{B}_j|^2 + \sum_{j\geq 1}|\mathcal{C}_j \cap G_2|^2) = O(|G_2|), \tag{7.10}$$

$$\mathbb{E}(\sum_{j\geq 1}|\tilde{\mathcal{A}}_j|^2 + \sum_{j\geq 1}|\tilde{\mathcal{C}}_j \cap G_1|^2) = O(|G_1|), \tag{7.11}$$

$$\mathbb{E}(\sum_{j\geq 1}|\tilde{\mathcal{B}}_j|^2 + \sum_{j\geq 1}|\tilde{\mathcal{C}}_j \cap G_2|^2) = O(|G_2|). \tag{7.12}$$

Now, we first assign spins to all components except the isolated vertices in $\{A_j\}$ and $\{\widetilde{A_j}\}$ independently in both chains. Let M_1, N_1 be the sum of spins in G_1 and G_2 respectively in first chain before assigning the rest of the spins, and similarly $\widetilde{M_1}$, $\widetilde{N_1}$ be the same for the second chain at this time. By (7.9),(7.10),(7.11),(7.12) and Markov's inequality that we have

$$\{\mathcal{A}, |M_1 - \widetilde{M_1}| = O(\sqrt{|G_1|}), |N_1 - \widetilde{N_1}| = O(\sqrt{|G_2|})\}$$

occurs with probability $\Omega(1)$. Then by Lemma 6.7, we can couple the sum of spins in both G_1 and G_2 so that they are the same in both chains with probability $\Omega(1)$. This gives the required coupling of (Y_1, Z_1) and $(\widetilde{Y_1}, \widetilde{Z_1})$. $\square$

CHAPTER 8

Critical Case

In this section, we prove that the mixing time for the Swendsen-Wang dynamics in the critical case $c = 2$ is of order $n^{1/4}$. This is part (ii) of Theorem 2.1.

Let X_t and Y_t be two magnetization chains such that X_t starts from an arbitrary location and Y_t starts from the stationary distribution. To prove an upper bound of order $n^{1/4}$ to the mixing time we show that we can couple X_t and Y_t so that they meet in time $O(n^{1/4})$ with probability $\Omega(1)$. For a high level view of this coupling strategy we refer the reader to Section 4.3.

Consider the following slight modification to the magnetization chain X_t. Instead of choosing a random spin for each component after the percolation step, we assign a positive spin to the largest component and random spins for all other components. Let X'_t be the sum of spins at time t (notice that we do *not* take absolute values here), that is,
(8.1)
$$X'_{t+1} \stackrel{d}{=} \max\{|\mathcal{C}_1^+(t)|, |\mathcal{C}_1^-(t)|\} + \epsilon \min\{|\mathcal{C}_1^+(t)|, |\mathcal{C}_1^-(t)|\} + \sum_{j \geq 2} \epsilon_j |\mathcal{C}_j^+(t)| + \sum_{j \geq 2} \epsilon'_j |\mathcal{C}_j^-(t)|,$$

where as usual ϵ, $\{\epsilon_j\}$ and $\{\epsilon'_j\}$ are independent mean zero $\pm$ signs. This chain has state space $[-n, n]$ and its absolute value is distributed as our original chain. As a consequence, any upper bound on the mixing time of the modified chain implies the same upper bound on the original chain.

The bulk of this section is devoted to the proof of the upper bound on the mixing time (the corresponding lower bound is much easier to prove and this is done in subsection 8.3). To ease the notation, in this section we will refer to this modified chain by X_t and Y_t. The only exception to this in this section is Theorem 8.24 where another modification to the chain was required for the proof.

The upper bound asserted in part (ii) of Theorem 2.1 will follow immediately by the following two theorems. Though their statement is almost identical, the difference in the starting point X_0 give rise to completely different proof methods so we chose to specify them as two separate theorems for convenience.

THEOREM 8.1. *Let X_t and Y_t be two SW magnetization chains such that $X_0 \geq n^{3/4}$ and $Y_0 \stackrel{d}{=} \pi$. Then we can couple X_t and Y_t so that they meet each other within $O(n^{1/4})$ steps with probability $\Omega(1)$.*

THEOREM 8.2. *Let X_t and Y_t be two SW magnetization chains such that $0 \leq X_0 \leq n^{3/4}$ and $Y_0 \stackrel{d}{=} \pi$. Then we can couple X_t and Y_t so that they meet each other within $O(n^{1/4})$ steps with probability $\Omega(1)$.*

Proof of the upper bound of part (ii) of Theorem 2.1: Theorem 8.1 and Theorem 8.2 give that for any $X_0 \geq 0$ we can couple X_t and Y_t so that they meet

within $O(n^{1/4})$ steps. If $X_0 < 0$, then by (8.1) and symmetry we have that
$$P(X_1 \geq 0) \geq \frac{1}{2},$$
so we may apply Theorem 8.1 and Theorem 8.2 again. This shows that the mixing time of X_t is bounded above by $O(n^{1/4})$. Note that $|X_t|$ and the original magnetization chain has the same distribution. Now Lemma 4.1 gives the required upper bound and concludes the proof. □

8.1. Starting at the $[n^{3/4}, n]$ regime: Proof of Theorem 8.1

THEOREM 8.3. [Crossing and overshoot] *Let X_t and Y_t be two SW magnetization chains with $X_0 \geq n^{3/4}$ and $Y_0 \stackrel{d}{=} \pi$. Put*
$$T = \min\left\{t : X_t, Y_t \in [A^{-1}n^{3/4}, An^{3/4}] \text{ and } |X_t - Y_t| \leq hn^{5/8}\right\},$$
for some constant $h > 0$ and large constant A. Then we can choose positive constants h, q, K depending only on A such that
$$\mathbf{P}(T \leq Kn^{1/4}) \geq q.$$

THEOREM 8.4. [Local CLT] *For any constants $A > 1$ and $h > 0$, there exist constants $\delta = \delta(A, h) > 0$ and $k = k(A, h) \in \mathbb{N}$ such that for any $x_0 \in [A^{-1}n^{3/4}, An^{3/4}]$ and any $x \in n + 2\mathbb{Z}$ with $|x - x_0| \leq hn^{5/8}$, we have*
$$\mathbf{P}(X_k = x | X_0 = x_0) \geq \delta n^{-5/8}.$$

Proof of Theorem 8.1: By Theorem 8.3, the event $T \leq Kn^{1/4}$ occurs with probability at least q. By Theorem 8.4 and the strong Markov Property we learn that there exist $\delta > 0$ and $k \in \mathbb{N}$ such that for any $x \in n+2\mathbb{Z}$ with $|x - X_T| \leq hn^{5/8}$ and $|x - Y_T| \leq hn^{5/8}$, we have
$$\mathbf{P}(X_{T+k} = x \mid T \leq Kn^{1/4}) \geq \delta n^{-5/8},$$
and
$$\mathbf{P}(Y_{T+k} = x \mid \tau \leq Kn^{1/4}) \geq \delta n^{-5/8}.$$
Thus, for any such x we can couple X_t and Y_t so that $X_{T+k} = Y_{T+k} = x$ with probability at least $\delta n^{-5/8}$. We have at least $\frac{hn^{5/8}}{2}$ such x's so in this coupling we have that $X_{T+k} = Y_{T+k}$ with probability at least $h\delta/2$. Lemma 3.2 concludes the proof. □

8.1.1. Crossing and overshoot: Proof of Theorem 8.3.
For any two magnetization chains X_t and Y_t, define $J_t = X_t - Y_t$. Let τ be the first time the two chains cross each other, i.e.

(8.2) $$\tau := \min\{t : \text{sign} J_t \neq \text{sign} J_0\}.$$

The following theorem implies Theorem 8.3 immediately.

THEOREM 8.5. *Let X_t and Y_t be two independent magnetization SW chain with $X_0 \geq n^{3/4}$ and $Y_0 \stackrel{d}{=} \pi$. There exists positive constants δ, K, A and h such that*
$$\mathbf{P}\left(\tau \leq Kn^{1/4}; \ X_{\tau-1}, Y_{\tau-1} \in [A^{-1}n^{3/4}, An^{3/4}]; J_{\tau-1} \leq hn^{5/8}\right) \geq \delta.$$

To prove Theorem 8.5 we will use the following results.

THEOREM 8.6. *The stationary distribution π of the modified magnetization chain satisfies*

$$\lim_{n\to\infty} \pi[a_1 n^{3/4}, a_2 n^{3/4}] = \frac{1}{Z}\int_{a_1}^{a_2} \exp(-\frac{1}{12}x^4)dx,$$

for any constants $a_2 \geq a_1 \geq 0$ where $Z = \int_0^\infty \exp(-\frac{1}{12}x^4)dx$ is the normalizing constant.

LEMMA 8.7. *For any constant $A > 0$ there exists N such that for all $n \geq N$ we have that if $X_0 \in [A^{-1}n^{3/4}, An^{3/4}]$, then the following hold:*

(i). $-Cn^{1/2} \leq \mathbb{E}X_1 - X_0 \leq 0$.
(ii). $\mathbb{E}|X_1 - x_0|^k \leq Cn^{5k/8}$ for $k = 2, 3, 4$.
(iii). $\mathbb{E}\sum_{j\geq 1}|\mathcal{C}_j^-|^2 \geq cn^{5/4}$.

where $C = C(A)$ and $c = c(A)$ are constants.

THEOREM 8.8. *Let X_t and Y_t be two independent magnetization chains with $X_0, Y_0 \in [b_1 n^{3/4}, b_2 n^{3/4}]$ for constants $b_2 > b_1 > 0$. Put $h = \frac{x_0 - y_0}{n^{5/8}}$ and suppose that $h > 0$ and that $h = o(n^{1/8})$. Let τ be the crossing time of X_t and Y_t defined in (8.2). Then there exist positive constants M and δ which only depend on b_1 and b_2 such that*

$$\mathbf{P}(\tau \leq Mh^2) \geq \delta.$$

LEMMA 8.9. *Let X_t be a magnetization SW chain and $I = [a_1 n^{3/4}, a_2 n^{3/4}]$ where $a_2 > a_1 > 0$ are two constants. Let $h \in (0, a_1)$ and $\xi \in [0, a_1/4]$ be two constants. Then for any $b \in I$, we have*

$$\mathbf{P}\big(\mathrm{sign}(X_1 - b) \neq \mathrm{sign}(X_0 - b) \,\big|\, X_0 > -\xi n^{3/4}, |X_0 - b| \geq hn^{3/4}\big) \leq Dn^{-1/3},$$

where $D = D(a_1, a_2, h, \xi)$ is a constant.

THEOREM 8.10. *For any fixed constants $b_2 > b_1 > 0$, $q < 1$ and $K > 0$, there exists a constant $B = B(b_1, b_2, q, K)$ such that for every $X_0 \in [b_1 n^{3/4}, b_2 n^{3/4}]$, we have*

$$\mathbf{P}\big(X_t \leq Bn^{3/4} \text{ for all } t \in [0, Kn^{1/4}]\big) \geq q.$$

THEOREM 8.11. *Let X_t be a magnetization SW chain with $X_0 > an^{3/4}$ where $a > 0$ is a constant. Define $\tau_a = \min\{t : X_t \leq an^{3/4}\}$. Then for any positive constant $b > 0$ we have*

(8.3) $$\mathbf{P}(\tau_a > bn^{1/4}) \leq \sqrt{\frac{6}{ab}}.$$

We begin by showing how these results imply the main theorem of this subsection.

Proof of Theorem 8.5: Let a_1, K and C be three positive constants to be selected later. Define

$$\tau_1 := \min\{t : X_t < a_1 n^{3/4}\},$$

and define $\mathcal{A}$ to be the event that

(1) $Y_0 \in [\frac{a}{2}n^{3/4}, an^{3/4}]$ and
(2) $\tau_1 \leq Kn^{1/4}$ and
(3) $Y_{\tau_1} \geq a_1 n^{3/4}$ and

(4) $Y_t \leq Cn^{3/4}$ for all $t \leq Kn^{1/4}$.

First we determine constants a_1, δ, K and C so that $\mathbf{P}(\mathcal{A}) \geq \delta > 0$. By Theorem 8.6, there exists a constant $q > 0$ such that
$$\mathbf{P}\Big(Y_0 \in [\frac{n^{3/4}}{2}, n^{3/4}]\Big) \geq q.$$
By Theorem 8.6 again, we can choose $a_1 > 0$ such that
$$\mathbf{P}\Big(Y_0 \in [-n, a_1 n^{3/4}]\Big) \leq \frac{q}{2}.$$
Since X_t and Y_t are independent we have that $Y_{\tau_1} \stackrel{d}{=} \pi_n$. Thus
$$\mathbf{P}\big(Y_0 \in [\frac{n^{3/4}}{2}, n^{3/4}], Y_{\tau_1} > a_1 n^{3/4}\big) \geq \frac{q}{2}.$$
By Lemma 8.11 there exists a constant $K = K(a_1, q)$ such that
$$(8.4) \qquad \mathbf{P}\big(Y_0 \in [\frac{n^{3/4}}{2}, n^{3/4}], Y_{\tau_1} > a_1 n^{3/4}, \tau_1 \leq Kn^{1/4}\big) \geq \frac{q}{4}.$$
By Lemma 8.10, there is a constant $C = C(K, q)$ such that
$$(8.5) \qquad \mathbf{P}\big(Y_t \leq Cn^{3/4} \text{ for all } t \leq Kn^{1/4}\big) \geq 1 - \frac{q}{8}.$$
Combining (8.4) and (8.5) shows that $\mathbf{P}(\mathcal{A}) \geq \frac{q}{8}$. Note that if $\mathcal{A}$ occurs, then
$$(8.6) \qquad \tau \leq \tau_1 \leq Kn^{1/4}.$$

Next we show $\{J_{\tau-1} \leq \frac{a_1}{2} n^{3/4}\} \cap \mathcal{A}$ has positive probability. We do this by proving $J_{\tau-1} \leq \frac{a_1}{2} n^{3/4}$ occurs with high probability on $\mathcal{A}$. Note that $\{J_{\tau-1} \leq \frac{a_1}{2} n^{3/4}\} \cap \mathcal{A}$ implies $X_{\tau-1}, Y_{\tau-1} \in [A^{-1} n^{3/4}, An^{3/4}]$ for some large constant A.

If $\{J_{\tau-1} > \frac{a_1}{2} n^{3/4}\} \cap \mathcal{A}$ occurs, then there exists some $t \leq Kn^{1/4}$ such that $J_t > \frac{a_1}{2} n^{3/4}$ and $J_{t+1} < 0$. This implies that there is a point $y \in [\frac{a_1}{2} n^{3/4}, (C + \frac{a_1}{2}) n^{3/4}]$ with $|X_t - y| \geq \frac{a_1}{4} n^{3/4}$ and $|Y_t - y| \geq \frac{a_1}{4} n^{3/4}$ and at least one of X_{t+1} and Y_{t+1} crosses y. Suppose first that $Y_{\tau-1} \geq -\xi n^{3/4}$ where ξ is a small positive constant. Then Lemma 8.9 and the union bound give that
$$(8.7) \qquad \mathbf{P}\Big(\mathcal{A}, J_{\tau-1} > \frac{a_1}{2} n^{3/4}, Y_{\tau-1} \geq -\xi n^{3/4}\Big) \leq Dn^{-1/3} Kn^{1/4} = o(1).$$

Next suppose that $Y_{\tau-1} < -\xi n^{3/4}$. Then there is a $t \in [0, Kn^{1/4}]$ such that $Y_t \leq -\xi n^{3/4}$. By (8.1), for any starting location, we have
$$\mathbf{P}(X_1 < -\xi n^{3/4}) \leq \mathbf{P}\Big(\epsilon \min\{|\mathcal{C}_1^+(t)|, |\mathcal{C}_1^-(t)|\} + \sum_{j \geq 2} \epsilon_j |\mathcal{C}_j^+(t)| + \sum_{j \geq 2} \epsilon_j' |\mathcal{C}_j^-(t)| < -\xi n^{3/4}\Big).$$
By Theorem 5.13 we have that
$$(8.8) \qquad \mathbb{E}\Big(\epsilon \min\{|\mathcal{C}_1^+(t)|, |\mathcal{C}_1^-(t)|\} + \sum_{j \geq 2} \epsilon_j |\mathcal{C}_j^+(t)| + \sum_{j \geq 2} \epsilon_j' |\mathcal{C}_j^-(t)|\Big)^4 = O(n^{8/3}),$$
so Markov's inequality gives that
$$(8.9) \qquad \mathbf{P}(X_1 < -\xi n^{3/4}) = O(n^{-1/3}).$$
The union bound implies now that
$$\mathbf{P}\Big(\mathcal{A}, J_{\tau-1} > \frac{a_1}{2} n^{3/4}, Y_{\tau-1} < -\xi n^{3/4}\Big) \leq O(n^{-1/3}) Kn^{1/4} = o(1),$$

8. CRITICAL CASE

and so together with (8.7) we conclude that $\{\mathcal{A}, X_{\tau-1}, Y_{\tau-1} \in [A^{-1}n^{3/4}, An^{3/4}]\}$ occurs with probability $\Omega(1)$ for some constant A. We denote this event by $\mathcal{B}$.

It remains to prove that $\{J_{\tau-1} \leq hn^{5/8}\} \cap \mathcal{B}$ occurs with probability $\Omega(1)$ for some constant $h > 0$. Suppose first $J_{\tau-1} > n^{23/32}$. Notice that $\{\mathcal{B}, J_{\tau-1} > n^{23/32}\}$ implies there is a $t < Kn^{1/4}$ such that $X_t, Y_t \in [A^{-1}n^{3/4}, An^{3/4}]$, $J_t > n^{23/32}$ and $J_{t+1} < 0$. This implies at least one of X_t and Y_t has to make a huge jump of order at least $n^{23/32}$. By part (ii) of Lemma 8.7 with $k = 4$, Markov's inequality and the union bound we have

$$(8.10) \qquad \mathbf{P}(J_{\tau-1} > n^{23/32}, \mathcal{B}) \leq O(n^{4(5/8-23/32)}n^{1/4}) = o(1).$$

To handle the case $J_{\tau-1} < n^{23/32}$, let

$$W_k = [2^k n^{5/8}, 2^{k+1} n^{5/8}],$$

and consider the probability $\mathbf{P}(J_{\tau-1} \in W_k, \mathcal{B})$. Let T_m be the first time that $J_t \in W_k$ for m-th time and

$$\mathcal{A}_m = \bigcap_{1 \leq m' \leq m} \{X_{T_{m'}}, Y_{T_{m'}} \in [A^{-1}n^{3/4}, A^{3/4}]\}.$$

Note that $\mathcal{A}_m \in \mathcal{F}_{T_m}$. We have

$$(8.11) \qquad \mathbf{P}(J_{\tau-1} \in W_k, \mathcal{B}) \leq \sum_{m=1}^{\infty} \mathbf{P}(T_m \leq \tau - 1, J_{T_m+1} < 0, \mathcal{B}).$$

Notice that $\{T_m \leq \tau - 1, J_{T_m+1} < 0, \mathcal{B}\}$ implies that for all $m' \leq m$, we have $X_{T_{m'}} > a_1 n^{3/4}$, $Y_{T_{m'}} < Cn^{3/4}$ and $|X_{T_{m'}} - Y_{T_{m'}}| \leq 2^{k+1} n^{5/8}$. This in particular implies that $X_{T_{m'}}, Y_{T_{m'}} \in [A^{-1}n^{3/4}, A^{3/4}]$. Hence $\{T_m \leq \tau - 1, J_{T_m+1} < 0, \mathcal{B}\}$ implies $\{T_m \leq \tau - 1, J_{T_m+1} < 0, \mathcal{A}_m\}$. Also, by part (ii) of Lemma 8.7 and Markov's inequality, we have

$$\mathbf{P}(|X_{t+1} - X_t| \geq 2^{k-1} n^{5/8} \,|\, X_t = \Theta(n^{3/4})) \leq \frac{C}{2^{4k}}.$$

The same inequality holds for Y_t by the same reason. Thus

$$\mathbf{P}\left(|J_{t+1} - J_t| \geq 2^k n^{5/8} \,\Big|\, X_t, Y_t = \Theta(n^{3/4})\right) \leq \frac{C}{2^{4k}}.$$

We now use the strong Markov property on the stopping time T_m and plug the above estimate in (8.11) to get that

$$(8.12) \qquad \begin{aligned} \mathbf{P}(J_{\tau-1} \in W_k, \mathcal{B}) &\leq \sum_{m=1}^{\infty} \mathbf{P}(T_m \leq \tau - 1, J_{T_m+1} < 0, \mathcal{A}_m) \\ &\leq \sum_{m=1}^{\infty} \frac{C}{2^{4k}} \mathbf{P}(T_m \leq \tau - 1, \mathcal{A}_m). \end{aligned}$$

If $\{T_m \leq \tau - 1, \mathcal{A}_m\}$ occurs, then for any $l \leq m$, we have $T_{m-l} \leq \tau - 1$ and $X_{T_{m-l}}, Y_{T_{m-l}} \in [A^{-1}n^{3/4}, An^{3/4}]$ and most importantly, the chains do not cross between time T_{m-l} and T_m, which is at least l steps. Now, let M and r be the constants from Lemma 8.8 and put $l = M2^{2k}$. The strong Markov property on the stopping time $T_{m-M2^{2k}}$ and Lemma 8.8 gives that

$$\mathbf{P}(T_m \leq \tau - 1, \mathcal{A}_m) \leq (1-r) \mathbf{P}(T_{m-M2^{2k}} \leq \tau - 1, \mathcal{A}_{m-M2^{2k}}).$$

Applying this recursively gives that
$$\mathbf{P}(T_m \leq \tau - 1, \mathcal{A}_m) \leq (1-r)^{[\frac{m}{M 2^{2k}}]}.$$
Plugging this into (8.12), we get
$$\mathbf{P}(J_{\tau-1} \in W_k, \mathcal{B}) \leq \frac{C}{2^{4k}} \sum_{m=1}^{\infty} (1-r)^{[\frac{m}{M 2^{2k}}]} = \frac{C}{2^{2k}}.$$

Combining this and (8.10) we have for large enough k_0, $\{J_{\tau-1} \leq 2^{k_0} n^{5/8}, \mathcal{B}\}$ occurs with probability $\Omega(1)$, which concludes the proof of the theorem. □

We now proceed with proving the statements we have used so far in the proof of Theorem 8.5. To prove Theorem 8.6 we will use the following small lemmas.

LEMMA 8.12 (Simon and Griffiths (1973)). *Denote by S_n the sum of spins for Ising model on the complete graph. If the inverse temperature $\beta = \frac{1}{n}$, then there exists a random variable X with density proportional to $\exp(-\frac{1}{12}x^4)$ such that*
$$\frac{S_n}{n^{3/4}} \xrightarrow{d} X,$$
as $n \to \infty$.

COROLLARY 8.13. *Consider Ising model on the complete graph with inverse temperature $\beta = \frac{1}{n} + O(\frac{1}{n^2})$. For any fixed constants $a_2 \geq a_1 \geq 0$, we have*
$$\lim_{n \to \infty} \mathbf{P}(|S_n| \in [a_1 n^{3/4}, a_2 n^{3/4}]) = \frac{1}{A} \int_{a_1}^{a_2} exp(-\frac{1}{12}x^4)dx,$$
where $A = \int_0^{\infty} exp(-\frac{1}{12}x^4)dx$ is the normalizing constant.

Proof of Corollary 8.13: By Lemma 8.12 we have that the conclusion of the corollary holds for $\beta_1 = \frac{1}{n}$. Thus it suffices to prove that for any configuration σ in which $|S_n(\sigma)| \in [a_1 n^{3/4}, a_2 n^{3/4}]$ we have
$$\mathbf{P}_\beta(\sigma) = (1+o(1))\mathbf{P}_{\beta_1}(\sigma).$$
Observe that on complete graph we have that
$$\sum_{u,v,u \neq v} \sigma(u)\sigma(v) = \frac{S_n^2 - n}{2}.$$
Thus, for any σ with $S_n(\sigma) \in [a_1 n^{3/4}, a_2 n^{3/4}]$, we have
$$(8.13) \quad \frac{\mathbf{P}_\beta(\sigma)}{\mathbf{P}_{\beta_1}(\sigma)} = \frac{e^{\beta(\frac{S_n^2-n}{2})}/Z(\beta)}{e^{\beta_1(\frac{S_n^2-n}{2})}/Z(\beta_1)} = (1+o(1))\frac{Z(\beta_1)}{Z(\beta)},$$
so it is enough to show $Z(\beta) = (1+o(1))Z(\beta_1)$. Indeed, Lemma 8.12 implies that
$$\mathbf{P}_{\beta_1}(|S_n| \geq n^{7/8}) = o(1),$$
but for any configuration σ with $|S_n(\sigma)| \leq n^{7/8}$ we have that
$$e^{\beta(\frac{S_n^2-n}{2})} = (1+o(1))e^{\beta_1(\frac{S_n^2-n}{2})},$$
and the assertion follows. □

LEMMA 8.14. *Let π_n be the stationary distribution of the modified magnetization SW chain X_t, then we have*

$$\lim_{n \to \infty} \pi_n[-\infty, 0] = 0.$$

Proof of Lemma 8.14: Recall that SW dynamics with parameter p has stationary distribution of Ising model with $p = 1 - e^{-2\beta}$. Plugging in $p = \frac{2}{n}$ we get $\beta = \frac{1}{n} + O(\frac{1}{n^2})$. By Corollary 8.13, for any $\epsilon > 0$, there exists constant b_1 and b_2 such that $0 < b_1 < b_2$ and

$$\pi_n\big([b_1 n^{3/4}, b_2 n^{3/4}] \cup [-b_2 n^{3/4}, -b_1 n^{3/4}]\big) > 1 - \epsilon.$$

By definition of stationarity, for any set S we have

(8.14) $$\sum_{y \in [-n,n]} \pi_n(y) \mathbf{P}(y, S) = \pi_n(S).$$

Put $S = [-n, 0]$ and denote $\pi_n[-n, 0]$ by δ_n. For any X_0 we have

$$\mathbf{P}(X_1 \le 0) \le 1/2,$$

by symmetry. For $X_0 \in [b_1 n^{3/4}, b_2 n^{3/4}]$ Lemma 8.9 gives that

$$P(X_1 < 0) \le D n^{-1/3}.$$

Plugging these into (8.14), we have

$$\delta_n = \sum_{y \in [-n,n]} \pi_n(y) \mathbf{P}(y, S) \le \frac{1}{2} \delta_n + \epsilon + D n^{-1/3},$$

which gives

$$\delta_n \le 2(\epsilon + D n^{-1/3}),$$

concluding the proof. □

Proof of Theorem 8.6: Directly follows from Corollary 8.13 and Lemma 8.14. □

The following is an easy estimate which use frequently to show that the main contribution from the first term of (8.1) comes from the $|\mathcal{C}_1^+|$ element rather than the $|\mathcal{C}_1^-|$ element.

PROPOSITION 8.15. *If $X_0 \ge C n^{2/3} \log^2 n$ for some large constant C, then*

$$\mathbf{P}(|\mathcal{C}_1^-| \ge |\mathcal{C}_1^+|) \le O(e^{-c \log^2 n}).$$

Proof. By our condition on X_0 we have that $|\mathcal{C}_1^+|$ is distributed as the size of the largest component in a supercritical random graph $G(m, p)$ with $m = \frac{n + X_0}{2}$ and $p = \frac{1+\epsilon}{m}$ with $\epsilon = X_0/n = \Omega(n^{-1/3} \log^2 n)$. Theorem 5.9 gives that

$$\mathbf{P}(|\mathcal{C}_1^+| \ge c n^{2/3} \log^2 n) \ge 1 - C e^{-c \log^2 n},$$

for some small $c > 0$. On the other hand $|\mathcal{C}_1^-|$ is distributed as a subcritical random graph. Theorem 1 of [25] gives that

$$\mathbf{P}(|\mathcal{C}_1^-| \ge c n^{2/3} \log^2 n) \le C e^{-c \log^2 n},$$

which finishes the proof. □

LEMMA 8.16. *If $X_t \geq Cn^{2/3} \log n$ for some large constant C, then*

$$\mathbb{E}[X_{t+1} \mid X_t] \leq X_t\left(1 - \frac{X_t}{6n}\right). \tag{8.15}$$

Proof. By (8.1) we have $\mathbb{E}[X_{t+1} \mid X_t] = \mathbb{E}[|\max\{|\mathcal{C}_1^+|, |\mathcal{C}_1^-|\}||X_t]$, hence Proposition 8.15 gives that

$$\mathbb{E}[X_{t+1} \mid X_t] = \mathbb{E}|\mathcal{C}_1^+| + O(e^{-c\log^2 n}).$$

Thus, Theorem 5.8 yields that

$$\mathbb{E}[X_{t+1} \mid X_t] \leq 2\frac{X_t}{n}\frac{n+X_t}{2} - \frac{7}{3}\frac{X_t^2}{n^2}\frac{n+X_t}{2} + O(e^{-c\log^2 n}) \leq X_t\left(1 - \frac{X_t}{6n}\right),$$

when n is large enough. $\square$

Proof of Lemma 8.7: As in the previous proof we have

$$\mathbb{E}X_1 = \mathbb{E}|\mathcal{C}_1^+| + O(e^{-c\log^2 n}).$$

Since $\epsilon = \frac{x_0}{n} = \Theta(n^{-1/4})$ Theorem 5.8 gives that

$$\begin{aligned}\mathbb{E}|\mathcal{C}_1^+| &= 2\frac{x_0}{n}\frac{n+x_0}{2} - \frac{8}{3}\left(\frac{x_0}{n}\right)^2\frac{n+x_0}{2} + O\left(\left(\frac{x_0}{n}\right)^3\frac{n+x_0}{2}\right) \\ &= x_0 - \frac{x_0^2}{3n} + O\left(\frac{x_0^3}{n^2}\right) = x_0 - \frac{x_0^2}{3n} + O(n^{1/4}),\end{aligned}$$

which gives part (i) of the lemma since $x_0 \in [A^{-1}n^{3/4}, An^{3/4}]$. We now prove part (ii). For $k = 2, 3, 4$, by (8.1) and Jensen's inequality we have that

$$\mathbb{E}|X_1 - x_0|^k = \mathbb{E}\Big||\mathcal{C}_1^+| - x_0 + \sum_{j\geq 2}\epsilon_j|\mathcal{C}_j^+| + \sum_{j\geq 1}\epsilon_j'|\mathcal{C}_j^-|\Big|^k + \Theta(e^{-cn^{1/8}})$$

$$\leq 2^{k-1}\left(\mathbb{E}\big||\mathcal{C}_1^+| - x_0\big|^k + \mathbb{E}\Big|\sum_{j\geq 2}\epsilon_j|\mathcal{C}_j^+| + \sum_{j\geq 1}\epsilon_j'|\mathcal{C}_j^-|\Big|^k\right). \tag{8.16}$$

Theorem 5.10 now gives that

$$\begin{aligned}\mathbb{E}\Big||\mathcal{C}_1^+| - 2\frac{x_0}{n}\frac{n+x_0}{2}\Big|^k &\leq C\left(\frac{n+x_0}{2}\Big/\frac{x_0}{n}\right)^{k/2} \\ &\leq C\left(\frac{n^2}{x_0}\right)^{k/2} \leq O(n^{5k/8}).\end{aligned}$$

Another application of Jensen's inequality gives that

$$\begin{aligned}\mathbb{E}\big||\mathcal{C}_1^+| - x_0\big|^k &= \mathbb{E}\Big|\Big(|\mathcal{C}_1^+| - 2\frac{x_0}{n}\frac{n+x_0}{2}\Big) + \Big(2\frac{x_0}{n}\frac{n+x_0}{2} - x_0\Big)\Big|^k \\ &\leq 2^{k-1}\left(O(n^{5k/8}) + \left(\frac{x_0^2}{n}\right)^k\right) \leq O(n^{5k/8}).\end{aligned} \tag{8.17}$$

To bound the rest of (8.16), notice that by Holder's inequality, we only need to consider the case $k = 4$. We have

$$\begin{aligned}\mathbb{E}\Big|\sum_{j\geq 2}\epsilon_j|\mathcal{C}_j^+| + \sum_{j\geq 1}\epsilon_j'|\mathcal{C}_j^-|\Big|^4 &\leq \sum_{j\geq 2}\mathbb{E}|\mathcal{C}_j^+|^4 + \sum_{j\geq 1}\mathbb{E}|\mathcal{C}_j^-|^4 + \Big(\sum_{j\geq 2}\mathbb{E}|\mathcal{C}_j^+|^2\Big)\Big(\sum_{j\geq 1}\mathbb{E}|\mathcal{C}_j^-|^2\Big) \\ &+ \mathbb{E}\sum_{i,j\geq 2, i\neq j}|\mathcal{C}_i^+|^2|\mathcal{C}_j^+|^2 + \mathbb{E}\sum_{i,j\geq 1, i\neq j}|\mathcal{C}_i^-|^2|\mathcal{C}_j^-|^2.\end{aligned}$$

By Theorem 5.10 we have
$$\sum_{j\geq 2}\mathbb{E}|\mathcal{C}_j^+|^2 \leq C_2 n\left(\frac{x_0}{n}\right)^{-1} = O(n^{5/4})$$
and
$$\sum_{j\geq 2}\mathbb{E}|\mathcal{C}_j^+|^4 \leq C_4 n\left(\frac{x_0}{n}\right)^{-5} = O(n^{9/4}).$$
By Theorem 5.12, we have
$$\sum_{j\geq 1}\mathbb{E}|\mathcal{C}_j^-|^2 \leq C_2 n\left(\frac{x_0}{n}\right)^{-1} = O(n^{5/4})$$
and
$$\sum_{j\geq 1}\mathbb{E}|\mathcal{C}_j^-|^4 \leq C_4 n\left(\frac{x_0}{n}\right)^{-5} = O(n^{9/4}).$$
These together with Theorem 5.13 to handle the cross terms finishes the proof of part (ii) of the lemma. Part (iii) follows immediately by Theorem 5.12,
$$\mathbb{E}\sum_{j\geq 1}|\mathcal{C}_j^-|^2 \geq c_2\frac{n-x_0}{2}\frac{n}{x_0} \geq c_2\frac{n}{4}A^{-1}n^{1/4} \geq cn^{5/4}.$$
□

LEMMA 8.17. *Let X be a real valued random variable with $\mathbb{E}X = 0$ and $\mathbb{E}X^2 \geq h^2$ and $\mathbb{E}X^4 \leq bh^4$ where $b \geq 1$. Then for any $\rho \in [0,1]$ we have*
$$\mathbf{P}(X \leq -\rho h) \geq \frac{(1-\rho^2)^2}{2b}.$$

Proof of Lemma 8.17: By Cauchy-Schwartz
$$\mathbb{E}[X^2 \mathbf{1}_{\{X^2 \geq \rho^2 h^2\}}] \leq \sqrt{\mathbb{E}X^4 \mathbb{E}\mathbf{1}_{\{X^2 \geq \rho^2 h^2\}}} \leq \sqrt{bh^4 \mathbf{P}(X^2 \geq \rho^2 h^2)}.$$
Hence,
$$h^2 \leq \mathbb{E}X^2 \leq \rho^2 h^2 + \mathbb{E}[X^2 \mathbf{1}_{\{X^2 \geq \rho^2 h^2\}}] \leq \rho^2 h^2 + \sqrt{bh^4 \mathbf{P}(X^2 \geq \rho^2 h^2)}.$$
We conclude that
$$\mathbf{P}(|X| \geq \rho h) \geq \frac{(1-\rho^2)^2}{b},$$
and the assertion follows by symmetry since $\mathbf{P}(X \leq -\rho h) = \mathbf{P}(-X \leq -\rho h)$. □

The following will be used in the proof of Theorem 8.8.

THEOREM 8.18. *Let X_t be a magnetization chain with $X_0 \in [b_1 n^{3/4}, b_2 n^{3/4}]$ where $b_2 > b_1 > 0$ are two constants. Let τ_1 be the first time that $X_t \notin [\frac{b_1}{2}n^{3/4}, (b_2 + \frac{b_1}{2})n^{3/4}]$. Then there exists a constant $C = C(b_1, b_2) > 0$ such that for all constant $\delta > 0$ we have*
$$\mathbf{P}(\tau_1 \leq \delta n^{1/4}) \leq C\delta^2.$$

Proof of Theorem 8.18 Denote by I the interval $[\frac{b_1}{2}n^{3/4}, (b_2 + \frac{b_1}{2})n^{3/4}]$. Part (ii) of Lemma 8.7 gives
(8.18) $$\mathbb{E}\left[(X_{(t+1)\wedge\tau_1} - X_{t\wedge\tau_1})^k \Big| \mathcal{F}_t\right] \leq Cn^{5k/8}$$

for $k = 2, 3, 4$. Define
$$Z := X_{(t+1)\wedge \tau_1} - X_{t\wedge \tau_1} - (\mathbb{E}X_{(t+1)\wedge \tau_1} - \mathbb{E}X_{t\wedge \tau_1}).$$
Note that $|\mathbb{E}X_{(t+1)\wedge \tau_1} - \mathbb{E}X_{t\wedge \tau_1}| \leq Cn^{1/2}$ by part (i) of Lemma 8.7, hence
$$\mathbb{E}\left[Z^k | \mathcal{F}_t\right] \leq Cn^{\frac{5k}{8}} \tag{8.19}$$
for $k = 2, 3, 4$. Also, for $k = 1$, part (i) of Lemma 8.7 gives that
$$\mathbb{E}[Z|\mathcal{F}_t] \leq C\sqrt{n}. \tag{8.20}$$
Denote
$$f(t) = \left(\mathbb{E}[X_{t\wedge \tau_1} - \mathbb{E}X_{t\wedge \tau_1}]^4\right)^{1/2}.$$
Note that
$$f(t+1)^2 = \mathbb{E}[X_{(t+1)\wedge \tau} - \mathbb{E}X_{(t+1)\wedge \tau_1}]^4 = \mathbb{E}\left[(X_{t\wedge \tau_1} - \mathbb{E}X_{t\wedge \tau_1}) + Z\right]^4. \tag{8.21}$$
For $k = 1, 2, 3, 4$, we have
$$\mathbb{E}\left[\left(X_{t\wedge \tau_1} - \mathbb{E}X_{t\wedge \tau_1}\right)^{4-k} Z^k\right] = \mathbb{E}\left(\mathbb{E}\left[\left(X_{t\wedge \tau_1} - \mathbb{E}X_{t\wedge \tau_1}\right)^{4-k} Z^k \Big| \mathcal{F}_t\right]\right)$$
$$= \mathbb{E}\left[\left(X_{t\wedge \tau_1} - \mathbb{E}X_{t\wedge \tau_1}\right)^{4-k} \mathbb{E}[Z^k|\mathcal{F}_t]\right].$$
Hölder's inequality implies that
$$\mathbb{E}\left[\left(X_{t\wedge \tau_1} - \mathbb{E}X_{t\wedge \tau_1}\right)^{4-k} Z^k\right] \leq Cn^{\frac{5k}{8}} f(t)^{\frac{4-k}{2}}, \tag{8.22}$$
for $k = 2, 3, 4$ and by (8.20)
$$\mathbb{E}\left[\left(X_{t\wedge \tau_1} - \mathbb{E}X_{t\wedge \tau_1}\right)^3 Z\right] \leq C\sqrt{n} f(t)^{\frac{3}{2}}. \tag{8.23}$$
Expanding the right hand side of (8.21) and plugging (8.22) and (8.23) into it, we get
$$f(t+1)^2 \leq f(t)^2 + C\sqrt{n} f(t)^{3/2} + Cn^{5/4} f(t) + Cn^{15/8} f(t)^{1/2} + Cn^{5/2}. \tag{8.24}$$
Comparing the right hand side of (8.24) with
$$\left(f(t) + Cn^{1/2} f(t)^{1/2} + Cn^{5/4}\right)^2, \tag{8.25}$$
we find that the first, second, third and fifth term of (8.24) is dominated by expanding (8.25). For the forth term, if $f(t) = O(n^{5/4})$, then it is dominated by $(Cn^{5/4})^2$. Otherwise it is dominated by $Cn^{5/4} f(t)$. The conclusion is that
$$f(t+1)^2 \leq \left(f(t) + Cn^{1/2} f(t)^{1/2} + Cn^{5/4}\right)^2. \tag{8.26}$$
Thus, if $f(t) = O(n^{3/2})$, then we have
$$f(t+1) \leq f(t) + Cn^{5/4}. \tag{8.27}$$
Since $f(0) = 0$, by iterating (8.27) we get that $f(t) \leq Ctn^{5/4}$ for all $t \leq \delta n^{1/4}$ where $\delta > 0$ is a constant. Put $t = \delta n^{1/4}$. Markov's inequality gives that
$$\mathbf{P}(|X_{\delta n^{1/4}\wedge \tau_1} - \mathbb{E}X_{\delta n^{1/4}\wedge \tau_1}| \geq \frac{b_1}{4} n^{3/4}) \leq \frac{(C\delta)^2}{\left(\frac{b_1}{4}\right)^4}. \tag{8.28}$$

By part (i) of Lemma 8.7 we have that
$$|\mathbb{E}X_{\delta n^{1/4} \wedge \tau_1} - X_0| \leq C\delta n^{3/4}.$$

Thus, for small enough δ we have
$$(8.29) \quad \mathbf{P}\left(\left|X_{\delta n^{1/4} \wedge \tau_1} - x_0\right| \leq \frac{b_1}{2}n^{3/4}\right) \geq 1 - C\delta^2,$$

which means that X_t has not jumped out of the window I within $\delta n^{1/4}$ steps with probability at least $1 - C\delta^2$. □

Proof of Theorem 8.8: Recall that $J_t = X_t - Y_t$. Let M be a large constant that will be chosen later. Assume without loss of generality that $J_0 \geq 0$, we will prove that J_{Mh^2} is negative with probability $\Omega(1)$, which implies the theorem. Denote by I the interval $[\frac{b_1}{2}n^{3/4}, (b_2 + \frac{b_1}{2})n^{3/4}]$ and define
$$\tau_1 = \min\{t : X_t \notin I \text{ or } Y_t \notin I\}.$$

We will prove our claim by precisely estimating the first, second and forth moment of $J_{Mh^2 \wedge \tau_1}$ and then apply Lemma 8.17 to $J_{Mh^2 \wedge \tau_1} - \mathbb{E}J_{Mh^2 \wedge \tau_1}$. We start with first moment estimate. By part (i) of Lemma 8.7 and the optional stopping theorem we get
$$(8.30) \quad \mathbb{E}X_{t \wedge \tau_1} - C\sqrt{n} \leq \mathbb{E}X_{(t+1) \wedge \tau_1} \leq \mathbb{E}X_{t \wedge \tau_1}$$

for some constant $C = C(b_1, b_2) > 0$. Applying (8.30) recursively gives that
$$(8.31) \quad X_0 - CMh^2\sqrt{n} \leq \mathbb{E}X_{Mh^2 \wedge \tau_1} \leq X_0.$$

The same formula holds for Y_t, hence
$$(8.32) \quad \mathbb{E}J_{Mh^2 \wedge \tau_1} \leq hn^{5/8} + CMh^2 n^{1/2}.$$

We proceed with the second moment estimate. Notice that if $X_0 \in I$, we have that
$$\mathbb{E}(X_1 - \mathbb{E}(X_1|\mathcal{F}_0))^2 = \mathbb{E}\Big[\max\{|\mathcal{C}_1^+|, |\mathcal{C}_1^-|\} - \mathbb{E}\max\{|\mathcal{C}_1^+|, |\mathcal{C}_1^-|\}$$
$$+ \epsilon \min\{|\mathcal{C}_1^+|, |\mathcal{C}_1^-|\} + \sum_{j \geq 2}\epsilon_j|\mathcal{C}_j^+| + \sum_{j \geq 2}\epsilon_j'|\mathcal{C}_j^-|\Big]^2$$

by (8.1). In the $n^{3/4}$ regime, we have $\mathbf{P}(|C_1^-| \geq |C_1^+|) = O(e^{-c\log^2 n})$ by Proposition 8.15, hence
$$\mathbb{E}(X_1 - \mathbb{E}(X_1|\mathcal{F}_0))^2 \geq (1 - Ce^{-c\log^2 n})\mathbb{E}\sum_{j \geq 1}|C_j^-|^2 \geq c_1 n^{5/4},$$

by part (iii) of Lemma (8.7). Also, by part (ii) of Lemma 8.7 we have that
$$\mathbb{E}(X_1 - \mathbb{E}(X_1|\mathcal{F}_0))^2 \leq Cn^{5/4}.$$

Now let
$$A_t = \sum_{i=0}^{t-1}(X_{(i+1) \wedge \tau_1} - X_{i \wedge \tau_1}) - \mathbb{E}(X_{(i+1) \wedge \tau_1} - X_{i \wedge \tau_1}|\mathcal{F}_i)$$

and
$$B_t = X_0 - \mathbb{E}X_{t \wedge \tau_1} + \sum_{i=0}^{t-1}\mathbb{E}(X_{(i+1) \wedge \tau_1} - X_{i \wedge \tau_1}|\mathcal{F}_i).$$

Then it is easy to verify that $A_t + B_t = X_{t \wedge \tau_1} - \mathbb{E} X_{t \wedge \tau_1}$. Moreover, since the martingale increments are orthogonal we have that

$$\mathbb{E} A_t^2 = \sum_{i=0}^{t-1} \mathbb{E}(X_{(i+1) \wedge \tau_1} - \mathbb{E}(X_{(i+1) \wedge \tau_1} | \mathcal{F}_i))^2.$$

Since $h = o(n^{1/8})$ Theorem 8.18 gives that

$$\mathbf{P}(\tau_1 \leq Mh^2) = o(1).$$

This implies that

$$cMh^2 n^{5/4} \leq \mathbb{E} A_{Mh^2}^2 \leq CMh^2 n^{5/4}.$$

By (8.31) and part (i) of Lemma 8.7, we get that $|B_t| \leq Ctn^{1/2}$. This gives

$$\mathbb{E} B_t^2 \leq Ct^2 n.$$

Cauchy-Schwarz inequality gives

$$\mathbb{E} |A_{Mh^2} B_{Mh^2}| \leq CMh^2 n^{9/8}.$$

Thus, we have

$$\text{Var} X_{Mh^2 \wedge \tau_1} = \mathbb{E} A_{Mh^2}^2 + \mathbb{E} B_{Mh^2}^2 + 2\mathbb{E} A_{Mh^2} B_{Mh^2} \geq (c - o(1)) Mh^2 n^{5/4}.$$

The same estimates hold for Y_t. Since X_t and Y_t are independent we have

(8.33) $$\text{Var} J_{Mh^2 \wedge \tau_1} \geq c_1 Mh^2 n^{5/4}.$$

For the fourth moment estimate, by (8.27) we have

$$\mathbb{E}[X_{Mh^2 \wedge \tau_1} - \mathbb{E} X_{Mh^2 \wedge \tau_1}]^4 \leq (Mh^2 Cn^{5/4})^2$$

and

$$\mathbb{E}[Y_{Mh^2 \wedge \tau_1} - \mathbb{E} Y_{Mh^2 \wedge \tau_1}]^4 \leq (Mh^2 Cn^{5/4})^2.$$

By the Jensen's inequality, we get

(8.34) $$\mathbb{E}[J_{Mh^2 \wedge \tau_1} - \mathbb{E} J_{Mh^2 \wedge \tau_1}]^4 \leq 16(Mh^2 Cn^{5/4})^2.$$

Putting (8.33) and (8.34) together, taking $\rho = \frac{1}{\sqrt{Mc_1}}$ and using Lemma 8.17, we get

$$\mathbf{P}(J_{Mh^2 \wedge \tau_1} - \mathbb{E} J_{Mh^2 \wedge \tau_1} \leq -hn^{5/8}) \geq \delta,$$

where $\delta > 0$ is a constant. Combining this with (8.32), we get

(8.35) $$\mathbf{P}\Big(J_{Mh^2 \wedge \tau_1} \leq 0\Big) \geq \delta.$$

Here we choose M so that $Mc_1 \geq 2$, concluding the proof. $\square$

Proof of Lemma 8.9: If $X_0 \leq b - hn^{3/4}$, then assume first $X_0 \in [\xi n^{3/4}, b - hn^{3/4}]$. In this regime, by part (ii) of Lemma 8.7 with $k = 4$ and Markov's inequality we have

$$\mathbf{P}\big(\text{sign}(X_1 - b) \neq \text{sign}(X_0 - b)\big) \leq \frac{Cn^{5/2}}{h^4 n^3} = O(n^{-1/2}).$$

Assume now $X_0 \in [-\xi n^{3/4}, \xi n^{3/4}]$. Recall the distribution of X_1 in (8.1) and that $b \geq a_1 n^{3/4}$ and $\xi \leq a_1/4$. If $X_1 \leq a_1 n^{3/4}$, then either

$$\max\{|\mathcal{C}_1^+|, |\mathcal{C}_1^-|\} > 2\xi n^{3/4},$$

or
$$\epsilon \min\{|\mathcal{C}_1^+(t)|, |\mathcal{C}_1^-(t)|\} + \sum_{j \geq 2} \epsilon_j |\mathcal{C}_j^+(t)| + \sum_{j \geq 2} \epsilon_j' |\mathcal{C}_j^-(t)| \geq \frac{a_1}{2} n^{3/4}.$$

By Theorem 5.9 and monotonicity of $|\mathcal{C}_1|$, we have
$$\mathbf{P}(\max\{|\mathcal{C}_1^+|, |\mathcal{C}_1^-|\} > 2\xi n^{3/4}) \leq C e^{-cn^{1/8}}.$$

By Theorem 5.13 and Markov's inequality, we have
$$\mathbf{P}(\epsilon \min\{|\mathcal{C}_1^+(t)|, |\mathcal{C}_1^-(t)|\} + \sum_{j \geq 2} \epsilon_j |\mathcal{C}_j^+(t)| + \sum_{j \geq 2} \epsilon_j' |\mathcal{C}_j^-(t)| \geq \frac{a_1}{2} n^{3/4}) = O(n^{-1/3}).$$

Thus, we have
$$\mathbf{P}(X_1 \geq b) = O(n^{-1/3}).$$

If $X_0 \geq b + hn^{3/4}$, then assume first $X_0 \in [b + h^{3/4}, Bn^{3/4}]$ for some large constant B. By part (ii) of Lemma 8.7 with $k = 4$ and Markov's inequality, we have
$$\mathbf{P}(X_1 \leq b) \leq \frac{Cn^{5/2}}{h^4 n^3} = O(n^{-1/2}).$$

Assume $X_0 \geq Bn^{3/4}$. If $X_1 \leq bn^{3/4}$, then either
$$\max\{|\mathcal{C}_1^+|, |\mathcal{C}_1^-|\} \leq \frac{B}{2} n^{3/4},$$

or
$$\epsilon \min\{|\mathcal{C}_1^+(t)|, |\mathcal{C}_1^-(t)|\} + \sum_{j \geq 2} \epsilon_j |\mathcal{C}_j^+(t)| + \sum_{j \geq 2} \epsilon_j' |\mathcal{C}_j^-(t)| \leq -(\frac{B}{2} - a_2) n^{3/4}.$$

By Theorem 5.9, we have
$$\mathbf{P}(\max\{|\mathcal{C}_1^+|, |\mathcal{C}_1^-|\} \leq \frac{B}{2} n^{3/4}) \leq C e^{-cn^{1/8}}.$$

By Theorem 5.13 and Markov's inequality we have
$$\mathbf{P}(\epsilon \min\{|\mathcal{C}_1^+(t)|, |\mathcal{C}_1^-(t)|\} + \sum_{j \geq 2} \epsilon_j |\mathcal{C}_j^+(t)| + \sum_{j \geq 2} \epsilon_j' |\mathcal{C}_j^-(t)| \leq O(n^{-1/3}).$$

Thus, we have
$$\mathbf{P}(X_1 \leq b) = O(n^{-1/3}).$$

$\square$

Proof of Theorem 8.10: Denote by I the interval $[b_1 n^{3/4}, b_2 n^{3/4}]$ and let B be a large constant to be chosen later. For any $X_0 \in I$, define
$$A_{t, X_0} = \mathbf{P}(X_t \text{ exceeds } Bn^{3/4} \text{ within } t \text{ steps} \mid X_0).$$

Let
$$A_t = \max_{X_0 \in I} A_{t, X_0}.$$

Then A_t is increasing in t. Let
$$\tau = \min\{t : X_t \notin [\frac{b_1}{2} n^{3/4}, Bn^{3/4}]\}.$$

Then $X_{t \wedge \tau}$ is a supermartingale by part (i) of Lemma 8.7. Thus we have

(8.36) $$\mathbf{E} X_{Kn^{1/4} \wedge \tau} \leq b_2 n^{3/4}.$$

For simplicity denote $g(B) = \max_{X_0 \in I} \mathbf{P}(X_{Kn^{1/4} \wedge \tau} \geq Bn^{3/4} \mid X_0)$. We get from the above estimate and (8.36) that $g(B) \to 0$ as $B \to \infty$. For all $X_0 \in I$ and $t \leq Kn^{1/4}$, we have

(8.37) $\quad A_{t,X_0} \leq g(B, X_0) + \mathbf{P}\Big(X_{Kn^{1/4} \wedge \tau} \leq \frac{b_1}{2} n^{3/4}, X_t \text{ exceeds } Bn^{3/4} \text{ before } t\Big).$

Denote
$$\mathcal{A} = \{X_{Kn^{1/4} \wedge \tau} \leq \frac{b_1}{2} n^{3/4}, X_t \text{ exceeds } Bn^{3/4} \text{ before } t\}.$$

Let τ_1 be the exit time of $[\frac{b_1}{2} n^{3/4}, (b_2 + \frac{b_1}{2}) n^{3/4}]$. By Theorem 8.18, we have
$$\mathbf{P}(\tau_1 > \delta n^{1/4}) \geq 1 - C\delta^2,$$
for any sufficiently small constant $\delta > 0$. On the event $\{\tau_1 > \delta n^{1/4}\}$, there are three cases:
  (i) $X_{\delta n^{1/4}} \in [\frac{b_1}{2} n^{3/4}, b_1 n^{3/4}]$.
  (ii) $X_{\delta n^{1/4}} \in [b_1 n^{3/4}, b_2 n^{3/4}]$.
  (iii) $X_{\delta n^{1/4}} \in [b_2 n^{3/4}, (b_2 + \frac{b_1}{2}) n^{3/4}]$.

For case (ii), by the Markov property at time $\delta n^{1/4}$, we have that
$$\mathbf{P}(\mathcal{A} \mid \tau_1 > \delta n^{1/4}, X_{\delta n^{1/4}} \in [b_1 n^{3/4}, b_2 n^{3/4}]) \leq A_{t - \delta n^{1/4}}.$$

For case (i), define
$$T = \min\{t > \delta n^{1/4} : X_t \in [b_1 n^{3/4}, b_2 n^{3/4}]\}.$$

By monotonicity of A_t and the strong Markov property on T we have
$$\mathbf{P}\big(\mathcal{A} \mid \tau_1 > \delta n^{1/4}, X_{\delta n^{1/4}} \in [\frac{b_1}{2} n^{3/4}, b_1 n^{3/4}], T < t\big) \leq A_{t - \delta n^{1/4}}.$$

The event $\{\mathcal{A}, \tau_1 > \delta n^{1/4}, X_{\delta n^{1/4}} \in [\frac{b_1}{2} n^{3/4}, b_1 n^{3/4}], T \geq t\}$ implies that there exists $t \leq Kn^{1/4}$ such that $X_t < \frac{b_1}{2} n^{3/4}$ and $X_{t+1} > b_2 n^{3/4}$. By Lemma 8.9 and the union bound, this happens with probability at most
$$Dn^{-\frac{1}{3}} Kn^{1/4} = O(n^{-1/12}).$$

For case (iii), the event $\{\mathcal{A}, \tau_1 > \delta n^{1/4}, X_{\delta n^{1/4}} \in [b_2 n^{3/4}, (b_2 + \frac{b_1}{2}) n^{3/4}]\}$ implies that X_t first goes below $\frac{b_1}{2} n^{3/4}$ and then goes above $Bn^{3/4}$. Let
$$T' = \min\{t : t > \tau, X_t \in I\}.$$

By monotonicity of A_t and the strong Markov property on T', we obtain
$$\mathbf{P}\big(\mathcal{A} \mid \tau_1 > \delta n^{1/4}, X_{\delta n^{1/4}} \in [b_2 n^{3/4}, (b_2 + \frac{b_1}{2}) n^{3/4}], T' < t\big) \leq A_{t - \delta n^{1/4}}.$$

By similar argument in case (ii), we have
$$\mathbf{P}\big(\mathcal{A}, \tau_1 > \delta n^{1/4}, X_{\delta n^{1/4}} \in [b_2 n^{3/4}, (b_2 + \frac{b_1}{2}) n^{3/4}], T' \geq t\big) = O(n^{-1/12}).$$

Summing up the above estimates, we obtain

(8.38) $\quad \mathbf{P}(\mathcal{A}, \tau_1 > \delta n^{1/4}) \leq A_{t - \delta n^{1/4}} + O(n^{-1/12}).$

On the event $\{\mathcal{A}, \tau_1 \leq \delta n^{1/4}\}$, which happens with probability at most $C\delta^2$, there are two cases to consider:
  (i) $X_{\tau_1} < \frac{b_1}{2} n^{3/4}$,
  (ii) $X_{\tau_1} > (b_2 + \frac{b_1}{2}) n^{3/4}$.

In case (i), let
$$T_1 = \min\{t : t > \tau_1, X_t \in I\}.$$
By monotonicity of A_t and the strong Markov property on T_1, we have
$$\mathbf{P}(\mathcal{A} \mid \tau_1 \leq \delta n^{1/4}, X_{\tau_1} < \frac{b_1}{2} n^{3/4}, T_1 < t) \leq A_t.$$
A similar argument as before gives us
$$\mathbf{P}(\mathcal{A}, \tau_1 \leq \delta n^{1/4}, X_{\tau_1} < \frac{b_1}{2} n^{3/4}, T_1 \geq t) = O(n^{-1/12}).$$
In case (ii), let
$$T_2 = \min\{t : t > \tau, X_t \in I\}.$$
Similar arguments gives
$$\mathbf{P}(\mathcal{A} \mid \tau_1 \leq \delta n^{1/4}, X_{\tau_1} < \frac{b_1}{2} n^{3/4}, T_2 < t) \leq A_t,$$
and
$$\mathbf{P}(\mathcal{A}, \tau_1 \leq \delta n^{1/4}, X_{\tau_1} < \frac{b_1}{2} n^{3/4}, T_2 \geq t) = O(n^{-1/12}).$$
Summing over these estimate, we obtain
$$(8.39) \qquad \mathbf{P}(\mathcal{A}, \tau_1 \leq \delta n^{1/4}) \leq C\delta^2 (A_t + O(n^{-1/12})).$$
Plugging (8.38) and (8.39) into (8.37), we get
$$A_{t,X_0} \leq g(B) + (A_{t-\delta n^{1/4}} + O(n^{-1/12})) + C\delta^2 (A_t + O(n^{-1/12})).$$
Maximizing over X_0 and rearranging give
$$A_t \leq \frac{1}{1 - C\delta^2} (A_{t-\delta n^{1/4}} + g(B) + O(n^{-1/12})).$$
Telescoping gives
$$A_{Kn^{1/4}} \leq \frac{1}{1 - C\delta^2}^{\left\lceil \frac{K}{\delta} \right\rceil} \left(C\delta^2 + g(B) + O(n^{-1/12}) \right).$$
Since $\frac{1}{1-C\delta^2}^{\left\lceil \frac{K}{\delta} \right\rceil}$ converges as δ goes to 0, we conclude that we can choose $\delta > 0$ small enough and B so large to make $A_{Kn^{1/4}}$ arbitrarily small, as required. $\square$

Proof of Theorem 8.11: Notice that
$$\begin{aligned} \mathbb{E}\Big[X_{(t+1)\wedge\tau_a} \Big| \mathcal{F}_t\Big] &= \mathbb{E}\Big[X_{t+1}\mathbf{1}_{\{\tau_a \geq t+1\}} + X_{\tau_a}\mathbf{1}_{\{\tau_a \leq t\}} \Big| \mathcal{F}_t\Big] \\ &= \mathbb{E}[X_{t+1}|\mathcal{F}_t]\mathbf{1}_{\{\tau_a \geq t+1\}} + X_{\tau_a}\mathbf{1}_{\{\tau_a \leq t\}}. \end{aligned}$$
By Lemma 8.16, we have
(8.40)
$$\mathbb{E}\Big[X_{(t+1)\wedge\tau_a}\Big|\mathcal{F}_t\Big] \leq X_t\Big(1 - \frac{X_t}{6n}\Big)\mathbf{1}_{\{\tau_a \geq t+1\}} + X_{\tau_a}\mathbf{1}_{\{\tau_a \leq t\}} = X_{t\wedge\tau_a} - \frac{X_t^2}{6n}\mathbf{1}_{\{\tau_a \geq t+1\}}.$$
Taking expectations on both sides of (8.40), we get
$$(8.41) \qquad \mathbb{E} X_{(t+1)\wedge\tau_a} \leq \mathbb{E} X_{t\wedge\tau_a} - \frac{1}{6n} \mathbb{E} X_t^2 \mathbf{1}_{\{\tau_a \geq t+1\}}.$$
Note that
$$\mathbb{E}\Big(X_t^2 \mathbf{1}_{\{\tau_a \geq t+1\}}\Big) \geq a^2 n^{3/2} \mathbf{P}(\tau_a \geq t+1),$$

and
$$a^2 n^{3/2} \geq \frac{\mathbb{E}\left(X_{\tau_a}^2 \mathbf{1}_{\{\tau_a \leq t\}}\right)}{\mathbf{P}(\tau_a \leq t)}.$$

Hence we have
$$\mathbb{E}\left(X_t^2 \mathbf{1}_{\{\tau_a \geq t+1\}}\right) \geq \frac{\mathbb{E}\left(X_{\tau_a}^2 \mathbf{1}_{\{\tau_a \leq t\}}\right)}{\mathbf{P}(\tau_a \leq t)} \mathbf{P}(\tau_a \geq t+1),$$

which implies
$$\mathbb{E}\left(X_{\tau_a}^2 \mathbf{1}_{\{\tau_a \leq t\}}\right) \leq \frac{\mathbf{P}(\tau_a \leq t)}{\mathbf{P}(\tau_a \geq t+1)} \mathbb{E}\left(X_t^2 \mathbf{1}_{\{\tau_a \geq t+1\}}\right).$$

Adding $\mathbb{E}\left(X_t^2 \mathbf{1}_{\{\tau_a \geq t+1\}}\right)$ to both sides, we obtain

(8.42) $$\frac{\mathbb{E}\left[X_t^2 \mathbf{1}_{\{\tau_a \geq t+1\}}\right]}{\mathbf{P}(\tau_a \geq t+1)} \geq \mathbb{E} X_{t \wedge \tau_a}^2 \geq (\mathbb{E} X_{t \wedge \tau_a})^2.$$

Plugging into (8.41), we get

(8.43) $$\mathbb{E} X_{(t+1) \wedge \tau_a} \leq \mathbb{E} X_{t \wedge \tau_a} - \frac{1}{6n} \mathbf{P}(\tau_a \geq t+1)(\mathbb{E} X_{t \wedge \tau_a})^2.$$

Note that $\mathbb{E} X_{(t+1) \wedge \tau_a} > 0$. Taking the inverse of (8.43) leads to
$$\frac{1}{\mathbb{E} X_{(t+1) \wedge \tau_a}} \geq \frac{1}{\mathbb{E} X_{t \wedge \tau_a}} + \frac{1}{6n} \mathbf{P}(\tau_a \geq t+1).$$

Summing t from 0 to $\lceil bn^{1/4} \rceil - 1$, we get

(8.44) $$\frac{1}{\mathbb{E} X_{\lceil bn^{1/4} \rceil \wedge \tau_a}} \geq \frac{1}{6n} \sum_{t=0}^{\lceil bn^{1/4} \rceil - 1} \mathbf{P}(\tau_a \geq t+1) \geq \frac{1}{6n} \mathbf{P}(\tau_a \geq bn^{1/4}) bn^{1/4}.$$

On the other hand, for any $x \in [0, n]$, observe that $X_{bn^{1/4} \wedge \tau_a} \leq -x$ implies there exists $t \leq bn^{1/4}$ such that $X_t > an^{3/4}$ and $X_{t+1} < -x$. This implies either
$$\max\{|\mathcal{C}_1^+|, |\mathcal{C}_1^-|\} \leq \frac{a}{2} n^{3/4},$$
or
$$\epsilon \min\{|\mathcal{C}_1^+(t)|, |\mathcal{C}_1^-(t)|\} + \sum_{j \geq 2} \epsilon_j |\mathcal{C}_j^+(t)| + \sum_{j \geq 2} \epsilon_j' |\mathcal{C}_j^-(t)| \leq -x - \frac{a}{2} n^{3/4}.$$

By Theorem 5.9, we have $\mathbf{P}(\max\{|\mathcal{C}_1^+|, |\mathcal{C}_1^-|\} \leq \frac{a}{2} n^{3/4}) = O(e^{-cn^{1/8}})$. By Theorem 5.13 and Markov's inequality, we have
$$\mathbf{P}(\epsilon \min\{|\mathcal{C}_1^+(t)|, |\mathcal{C}_1^-(t)|\} + \sum_{j \geq 2} \epsilon_j |\mathcal{C}_j^+(t)| + \sum_{j \geq 2} \epsilon_j' |\mathcal{C}_j^-(t)| \leq -x - \frac{a}{2} n^{3/4})) \leq \frac{Cn^{8/3}}{(x + \frac{a}{2} n^{3/4})^4}.$$

Hence by union bound we obtain
$$\mathbf{P}(X_{bn^{1/4} \wedge \tau_a} \leq -x) \leq \frac{Cn^{8/3}}{(x + \frac{a}{2} n^{3/4})^4} bn^{1/4}.$$

By a direct computation we obtain
$$\mathbb{E}(|X_{bn^{1/4} \wedge \tau_a}| \mathbf{1}_{\{X_{bn^{1/4} \wedge \tau_a} \leq 0\}}) \leq \sum_{x=0}^{n} bn^{1/4} \frac{Cn^{8/3}}{(x + \frac{a}{2} n^{3/4})^4} = O(n^{2/3}).$$

Thus we get
$$\mathbb{E} X_{\lceil bn^{1/4}\rceil \wedge \tau_a} \geq \mathbf{P}(\tau_a > \lceil bn^{1/4}\rceil) an^{3/4} - O(n^{2/3}).$$
Multiplying this and (8.44) we get
$$1 \geq \frac{1}{6n} bn^{1/4} an^{3/4} \Big[\mathbf{P}(\tau_a > \lceil bn^{1/4}\rceil)\Big]^2,$$
which gives (8.3). □

8.1.2. Coupling inside the scaling window: Proof of Theorem 8.4.

LEMMA 8.19. *For any fixed constant $A > 1$ there exist positive constants $q = q(A), \beta = \beta(A)$, such that if $X_0 \in [A^{-1}n^{3/4}, An^{3/4}]$, then*
$$\mathbf{P}(X_1 = x | X_0) \geq q n^{-5/8}$$
for any $x \in n + 2\mathbb{Z}$ and $|x - X_0| \leq \beta n^{5/8}$.

Proof of Theorem 8.4 We will use induction to prove that for any $\ell > 0$ and any $x \in n + 2\mathbb{Z}$ such that $|X_0 - x| \leq \beta(1/2 + \ell/2)n^{5/8}$, we have

(8.45) $$\mathbf{P}(X_\ell = x | X_0) \geq q^\ell \left(\frac{\beta}{2}\right)^{\ell-1} n^{-5/8}.$$

This implies Theorem 8.4 immediately.

We prove this assertion by induction on ℓ. Lemma 8.19 implies (8.45) is true for $\ell = 1$. Suppose now (8.45) holds for ℓ and we prove for $\ell + 1$. If $x \in n + 2\mathbb{Z}$ and $|x - X_0| \leq \beta\big(1/2 + (\ell+1)/2\big)n^{5/8}$, then the number of y such that $y \in n + 2\mathbb{Z}$ and $|y - x| \leq \beta n^{5/8}$ and $|y - X_0| \leq \beta(1/2 + \ell/2)n^{5/8}$ is at least $\frac{\beta}{2} n^{5/8}$. Thus, we have

$$\mathbf{P}\Big(|X_\ell - x| \leq \beta n^{5/8}\Big) = \sum_{|y-x| \leq \beta n^{5/8}} \mathbf{P}(X_\ell = y)$$

(8.46) $$\geq q^\ell \left(\frac{\beta}{2}\right)^{\ell-1} n^{-5/8} \frac{\beta}{2} n^{5/8} = q^\ell \left(\frac{\beta}{2}\right)^\ell,$$

where we used the induction hypothesis. Since $|x - X_0| \leq \beta\big(1/2 + (\ell+1)/2\big)n^{5/8}$, we get
$$\mathbf{P}\Big(X_{\ell+1} = x \,\Big|\, |X_\ell - x| \leq \beta n^{5/8}\Big) \geq q n^{-5/8}$$
by Lemma 8.19. Together with (8.46) we get (8.45) for $\ell + 1$, concluding the proof. □

Recall that conditioned on the cluster sizes, X_1 is a summation of independent but not identically distributed random variables. The following is a local central limit theorem for such sums, tailored to our particular needs, and is used to prove Lemma 8.19. We have not found in the literature a statement general enough to be valid in our setting. The proof is the standard proof of the local CLT using characteristic function.

LEMMA 8.20. *Suppose K_n are positive integers such that $K_n \geq qn$ for some constant $q > 0$ and $a_1, a_2, \cdots, a_{K_n}$ are positive integers such that $a_j = 1$ for $1 \leq j \leq qn$ and $a_j \leq \sqrt{\frac{qn}{2}}$ for all j. Let $b(n) = \sum_{j=1}^{K_n} a_j$ and $c(n) = \sqrt{\sum_{j=1}^{K_n} a_j^2 / n^{5/4}}$. Assume that there are two positive constants δ and C such that $\delta < c(n) < C$ for* all n. Let $X_n = \sum_{j=1}^{K_n} \epsilon_j a_j$ where $\{\epsilon_j\}$ is independent random $\pm$ signs. Then for any $x \in b(n) + 2\mathbb{Z}$ and large enough n, we have

(8.47) $$\mathbf{P}\Big(X_n = x\Big) \geq \frac{\sqrt{2}}{\sqrt{\pi}c(n)n^{5/8}}\Big(e^{-\frac{x^2}{2}} - 1/2\sqrt{2}\Big).$$

Proof of Lemma 8.19: We need to show that with probability $\Omega(1)$ the percolation configuration fits the setting of Lemma 8.20. Define $\mathcal{A}_1$ and $\mathcal{A}_2$ as the following events:

$$\mathcal{A}_1 = \{|\mathcal{C}_1^+| \in [X_0 - \frac{c}{4}n^{5/8}, X_0 + \frac{c}{4}n^{5/8}], \sum_{j \geq 2}|\mathcal{C}_j^+|^2 \leq Dn^{5/4}, \sum_{j \geq 2}|\mathcal{C}_j^+|^3 \leq Dn^{7/4}\},$$

$$\mathcal{A}_2 = \{\sum_{j \geq 1}|\mathcal{C}_j^-|^2 \leq Dn^{5/4}, \sum_{j \geq 1}|\mathcal{C}_j^-|^3 \leq Dn^{7/4}, \sum_{|\mathcal{C}_j^-| \leq \frac{\sqrt{n}}{6}}|\mathcal{C}_j^-|^2 \geq c^2 n^{5/4}, |\{j : |\mathcal{C}_j^-| = 1\}| \geq \frac{n}{18}\}$$

where D and c are constants to be selected later. First we prove that $\mathcal{A}_1$ and $\mathcal{A}_2$ both happen with probability $\Omega(1)$. To bound from below the probability of $\mathcal{A}_2$, take $\delta = \frac{1}{3\sqrt{2}}$ in Theorem 5.14. We get

(8.48) $$\mathbf{P}\Big(\sum_{|\mathcal{C}_j^-| \leq \frac{1}{6}\sqrt{n}}|\mathcal{C}_j^-|^2 \geq cn^{5/4}\Big) \geq q = q(A) > 0,$$

for some $c = c(A) > 0$. By Theorem 5.12, for $k = 2, 3$ we have

$$\mathbb{E}\sum_{j \geq 1}|\mathcal{C}_j^-|^k \leq Cn(A^{-1}n^{-1/4})^{-2k+3}.$$

Thus, for

(8.49) $$D \geq \frac{4CA^3}{q},$$

we have by Markov's inequality that

(8.50) $$\mathbf{P}\Big(\sum_{j \geq 1}|\mathcal{C}_j^-|^2 \geq Dn^{5/4}\Big) \leq \frac{q}{4}$$

and

(8.51) $$\mathbf{P}\Big(\sum_{j \geq 1}|\mathcal{C}_j^-|^3 \geq Dn^{7/4}\Big) \leq \frac{q}{4}.$$

By Lemma 5.7, we have

(8.52) $$\mathbf{P}\Big(|\{j : |\mathcal{C}_j^-| = 1\}| \geq \frac{n}{18}\Big) \geq 1 - C/n \geq 1 - \frac{q}{4}.$$

Putting (8.48), (8.50), (8.51) and (8.52) together, we get

$$\mathbf{P}(\mathcal{A}_2) \geq \frac{q}{4}.$$

To bound from below the probability of $\mathcal{A}_1$, we apply Theorem 5.11 to get that

$$\mathbf{P}\Big(\big||\mathcal{C}_1^+| - x_0(1 + \frac{x_0}{n})\big| \leq \frac{c}{4}(\frac{n+x_0}{2})^{5/8}\Big) \geq q = q(A) > 0.$$

Since $\frac{x_0^2}{n} = o(n^{5/8})$ and $\frac{n+x_0}{2} \leq \frac{3}{4}n$, we get

(8.53) $$\mathbf{P}\Big(|\mathcal{C}_1^+| \in \Big[x_0 - \frac{c}{4}n^{5/8}, x_0 + \frac{c}{4}n^{5/8}\Big]\Big) \geq q.$$

8. CRITICAL CASE

By Theorem 5.13, for $k = 2, 3$ we have
$$\mathbb{E}\sum_{j\geq 2}|\mathcal{C}_j^+|^k \leq C_k n(A^{-1}n^{-1/4})^{-2k+3}.$$

Again, when D satisfies 8.49 we get by Markov's inequality that

(8.54) $$\mathbf{P}\Big(\sum_{j\geq 2}|\mathcal{C}_j^+|^2 \geq Dn^{5/4}\Big) \leq \frac{q}{4}$$

and

(8.55) $$\mathbf{P}\Big(\sum_{j\geq 2}|\mathcal{C}_j^+|^3 \geq Dn^{7/4}\Big) \leq \frac{q}{4}.$$

By (8.53), (8.54) and (8.55), we have

(8.56) $$\mathbf{P}(\mathcal{A}_1) \geq \frac{q}{2}.$$

Since $\mathcal{A}_1$ and $\mathcal{A}_2$ are independent, we get
$$\mathbf{P}(\mathcal{A}_1 \cap \mathcal{A}_2) \geq \frac{q^2}{8},$$
providing D satisfies (8.49). By Proposition 8.15 we have $\mathbf{P}(|C_1^-| \geq |C_1^+|) = O(e^{-c\log^2 n})$. Hence the event
$$\mathcal{A} = \{\mathcal{A}_1, \mathcal{A}_2, |C_1^-| < |C_1^+|)\},$$
occurs with probability $\Omega(1)$.

Next we prove that for every $x \in n + 2\mathbb{Z}$ and $|x - X_0| \leq \frac{c}{2}n^{5/8}$, there exist a constant $\delta > 0$ such that

(8.57) $$\mathbf{P}(X_1 = x \mid \mathcal{A}) \geq \delta n^{-5/8},$$

which will conclude the proof. Denote
$$M_1 = |\mathcal{C}_1^+| + \sum_{j\geq 2}\epsilon_j|\mathcal{C}_j^+| + \sum_{|\mathcal{C}_j|>\sqrt{n}/6}\epsilon_j'|\mathcal{C}_j^-|$$

and
$$M_2 = \sum_{|\mathcal{C}_j|\leq\sqrt{n}/6}\epsilon_j'|\mathcal{C}_j^-|.$$

Note that M_1 and M_2 are independent conditioned on $\mathcal{A}$. We will first prove that there exist a constant $\alpha > 0$ such that

(8.58) $$\mathbf{P}\Big(|M_1 - X_0| \leq \frac{c}{2}n^{5/8}\Big|\mathcal{A}\Big) \geq \alpha.$$

On $\mathcal{A}$ we have that

(8.59) $$\sum_{j\geq 2}|\mathcal{C}_j^+|^2 + \sum_{|\mathcal{C}_j|>\sqrt{n}/6}|\mathcal{C}_j^-|^2 \leq 2Dn^{5/4}$$

and

(8.60) $$\sum_{j\geq 2}|\mathcal{C}_j^+|^3 + \sum_{|\mathcal{C}_j|>\sqrt{n}/6}|\mathcal{C}_j^-|^3 \leq 2Dn^{7/4}.$$

If $\sum_{j\geq 2}|\mathcal{C}_j^+|^2 + \sum_{|\mathcal{C}_j|>\sqrt{n}/6}|\mathcal{C}_j^-|^2 \leq \frac{c^2}{32}n^{5/4}$, by Markov's inequality, we have

(8.61) $$\mathbf{P}\Big(\Big|\sum_{j\geq 2}\epsilon_j|\mathcal{C}_j^+|\Big| + \sum_{|\mathcal{C}_j|>\sqrt{n}/6}\epsilon_j'|\mathcal{C}_j^-|\Big| \leq \frac{c}{4}n^{5/8}\Big) \geq 1/2.$$

Otherwise
$$\frac{|\mathcal{C}_j|}{\left(\sum_{j\geq 2}|\mathcal{C}_j^+|^2 + \sum_{|\mathcal{C}_j|>\sqrt{n}/6}|\mathcal{C}_j^-|^2\right)^{1/2}} > \epsilon$$
implies
$$|\mathcal{C}_j| \geq \frac{\epsilon c^2}{32} n^{5/8},$$
and since $\{|\mathcal{C}_j|\}$ also satisfy (8.60), we learn that the Lindeberg condition is satisfied. By Lindeberg-Feller theorem (see [**10**], (4.5)), we have

(8.62) $$\mathbf{P}\Big(\Big|\sum_{j\geq 2}\epsilon_j|\mathcal{C}_j^+| + \sum_{|\mathcal{C}_j|>\sqrt{n}/6}\epsilon_j'|\mathcal{C}_j^-|\Big| \leq \frac{c}{4}n^{5/8}\Big) \geq \alpha > 0.$$

Combining this and (8.61) yields (8.58).

To estimate M_2 let
$$b = \sum_{|\mathcal{C}_j|\leq\sqrt{n}/6}|\mathcal{C}_j^-| \quad \text{and} \quad a = n^{-9/8}b^{1/2}.$$

By Lemma 8.20, for every $x \in b + 2\mathbb{Z}$, we have
$$\mathbf{P}(M_2 = x|\mathcal{A}) \geq \frac{\sqrt{2}}{\sqrt{\pi}an^{5/8}}\Big(e^{-\frac{x^2}{2a^2n^{5/4}}} - 1/2\Big).$$

For all x such that $|x| \leq cn^{5/8}$, we have
$$\frac{\sqrt{2}}{\sqrt{\pi}an^{5/8}}\Big(e^{-\frac{x^2}{2a^2n^{5/4}}} - 1/2\Big) \geq \frac{\sqrt{2}}{\sqrt{\pi}Dn^{5/8}}\Big(e^{-1/2} - 1/2\Big) \geq \delta n^{-5/8}$$
where δ is a constant. So for every $x \in b + 2\mathbb{Z}$ and $|x| \leq cn^{5/8}$, we have

(8.63) $$\mathbf{P}(M_2 = x|\mathcal{A}) \geq \delta n^{-5/8}.$$

By (8.58) and (8.63), for every $x \in n + 2\mathbb{Z}$ with $|x - x_0| \leq \frac{c_1}{2}n^{5/8}$, we have
$$\mathbf{P}(M_1 + M_2 = x|\mathcal{A})$$
$$\geq \mathbf{P}\Big(|M_1 - x_0| \leq \frac{c_1}{2}n^{5/8}, M_2 = (x - x_0) - (M_1 - x_0)\Big|\mathcal{A}\Big)$$
$$\geq \alpha\delta n^{-5/8}.$$

This proves (8.57), which concludes the whole proof. $\square$

To prove Lemma 8.20 we need the following two small assertions. The first is Exercise 3.2 of [**10**].

LEMMA 8.21. *If $\mathbf{P}(X \in b + h\mathbb{Z}) = 1$, where b is a complex number and $h > 0$ is a real number. Then for any $x \in b + h\mathbb{Z}$, we have*
$$\mathbf{P}(X = x) = \frac{h}{2\pi}\int_{-\pi/h}^{\pi/h} e^{-itx}\phi(t)dt,$$
where $\phi(t)$ is the characteristic function of X.

LEMMA 8.22. *For any x in $\mathbb{R}$, let $m(x)$ be the integer that is closest to x (if $x - \frac{1}{2}$ is an integer, then we put $m(x) = x - \frac{1}{2}$). Then for any x*
$$|\cos x| \leq \exp\Big(-\frac{(x - m(\frac{x}{\pi})\pi)^2}{2}\Big).$$

8. CRITICAL CASE

Proof. Since $m(\frac{x}{\pi})\pi \in \{k\pi\}_{k\in\mathbb{Z}}$, we have $|\cos x| = |\cos(x - m(\frac{x}{\pi})\pi)|$. Also, we have $-\frac{\pi}{2} \leq x - m(\frac{x}{\pi})\pi \leq \frac{\pi}{2}$. Since $\cos x \leq e^{-\frac{x^2}{2}}$ for all $x \in [-\frac{\pi}{2}, \frac{\pi}{2}]$ we have that

$$|\cos x| = \cos\left(x - m(\frac{x}{\pi})\pi\right) \leq \exp\left(-\frac{(x - m(\frac{x}{\pi})\pi)^2}{2}\right).$$

□

Proof of Lemma 8.20: For simplicity we will abbreviate $c(n)$ by c. Let

$$d_j = \frac{a_j}{cn^{5/8}}.$$

Then we have $\sum_{j=1}^{K_n} d_j^2 = 1$ and $\frac{X_n}{cn^{5/8}} = \sum_{j=1}^{K_n} \epsilon_j d_j$. Since $a_j = O(n^{1/2})$, we have that $d_j = O(n^{-1/8})$. Thus it satisfies Lindeberg condition (see [**10**]). Consequently, we have that

$$\frac{X_n}{cn^{5/8}} \xrightarrow{d} N(0,1).$$

Denote the characteristic function of $\frac{X_n}{cn^{5/8}}$ by $\phi_n(t)$. A straightforward computation gives that

$$(8.64) \qquad \phi_n(t) = \left(\cos\frac{t}{cn^{5/8}}\right)^{qn} \prod_{j=qn+1}^{K_n} \cos(td_j),$$

and we have $\phi_n(t) \to e^{-\frac{t^2}{2}}$ for all fixed $t \in \mathbb{R}$. Taking $h = \frac{2}{cn^{5/8}}$ in Lemma 8.21, for $x \in b(n) + 2\mathbb{Z}$ we have

$$(8.65) \qquad \mathbf{P}(X_n = x) = \frac{1}{\pi cn^{5/8}} \int_{-\frac{\pi}{2}cn^{5/8}}^{\frac{\pi}{2}cn^{5/8}} e^{-itx}\phi_n(t)\mathrm{d}t.$$

Let M be a large constant to be selected later. Note that $\phi_n(t)$ is an even function so

$$\int_{-\frac{\pi}{2}cn^{5/8}}^{\frac{\pi}{2}cn^{5/8}} e^{-itx}\phi_n(t)\mathrm{d}t = \int_{-M}^{M} e^{-itx}\phi_n(t)\mathrm{d}t + 2\int_{M}^{\frac{\pi}{2}cn^{5/8}} e^{-itx}\phi_n(t)\mathrm{d}t$$

$$(8.66) \qquad \geq \int_{-M}^{M} e^{-itx}\phi_n(t)\mathrm{d}t - 2\int_{M}^{\frac{\pi}{2}cn^{5/8}} |\phi_n(t)|\mathrm{d}t.$$

We will first bound from above the second term of (8.66). Let $m_j(t) = m\left(\frac{td_j}{\pi}\right)\frac{\pi}{d_j}$, i.e., $m_j(t)$ is the element in $\left\{k\frac{\pi}{d_j}\right\}_{k\in\mathbb{Z}}$ that is closest to t. Note that by Lemma 8.22, we have

$$\cos(td_j) \leq \exp\left\{-\left[td_j - m\left(\frac{td_j}{\pi}\right)\pi\right]^2 \Big/ 2\right\} = \exp\left\{-d_j^2 \frac{(t - m_j(t))^2}{2}\right\}.$$

For large enough n, we have $\frac{1}{c^2 n^{5/4}} \geq \frac{1}{2}\frac{1}{c^2 n^{5/4} - qn}$. Thus, we get

$$\left|\cos(td_j)\right| \leq \exp\left\{-\frac{a_j^2}{c^2 n^{5/4} - qn} \cdot \frac{(t - m_j(t))^2}{4}\right\}.$$

Since $\sum_{j=qn+1}^{K_n} \frac{(a_j)^2}{c^2 n^{5/4} - qn} = 1$ and e^{-x} is a convex function, we have by Jensen's inequality that

$$\prod_{j=qn+1}^{K_n} |\cos(td_j)| \leq \exp\left\{ - \sum_{j=qn+1}^{K_n} \frac{a_j^2}{c^2 n^{5/4} - qn} \frac{(t-m_j(t))^2}{4} \right\}$$

(8.67)
$$\leq \sum_{j=qn+1}^{K_n} \frac{a_j^2}{c^2 n^{5/4} - qn} \exp\left(- \frac{(t-m_j(t))^2}{4} \right).$$

Recall that $|t| \leq \frac{\pi}{2} cn^{5/8}$ and $|\cos(x)| \leq e^{-\frac{x^2}{2}}$ for $x \in [-\frac{\pi}{2}, \frac{\pi}{2}]$, whence

(8.68)
$$\left| \cos \frac{t}{cn^{5/8}} \right|^{qn} \leq \exp\left(- \frac{qt^2}{2c^2 n^{1/4}} \right).$$

Plugging (8.67) and (8.68) into (8.64), we get

$$|\phi_n(t)| \leq \sum_{j=qn+1}^{K_n} \frac{a_j^2}{c^2 n^{5/4} - qn} \exp\left(- \frac{(t-m_j(t))^2}{4} - \frac{qt^2}{2c^2 n^{1/4}} \right).$$

Hence, we have
(8.69)
$$\int_M^{\frac{\pi}{2} cn^{5/8}} |\phi_n(t)| dt \leq \sum_{j=qn+1}^{K_n} \frac{a_j^2}{c^2 n^{5/4} - qn} \int_M^{\infty} \exp\left(- \frac{(t-m_j(t))^2}{4} - \frac{qt^2}{2c^2 n^{1/4}} \right) dt.$$

We will divide the integral into two parts such that the first part converges to 0 as M goes to infinity and the second part is bounded by a constant. Recall that $m_j(t) = 0$ for $t \in [-\frac{\pi}{2d_j}, \frac{\pi}{2d_j}]$, so for any $j \in [qn+1, K_n]$, we have

$$\int_M^{\infty} \exp\left(- \frac{(t-m_j(t))^2}{4} - \frac{qt^2}{2c^2 n^{1/4}} \right) dt$$

$$= \int_M^{\frac{\pi}{2d_j}} \exp\left(- \frac{t^2}{4} - \frac{qt^2}{2c^2 n^{1/4}} \right) dt$$

(8.70)
$$+ \sum_{\ell=1}^{\infty} \int_{\frac{\pi}{2d_j}(2\ell-1)}^{\frac{\pi}{2d_j}(2\ell+1)} \exp\left(- \frac{(t-m_j(t))^2}{4} - \frac{qt^2}{2c^2 n^{1/4}} \right) dt$$

The first term of the right hand side of (8.70) is bounded by $\int_M^{\infty} e^{-\frac{t^2}{4}} dt$. For the second term, note that for $t \geq \frac{\pi}{2d_j}(2\ell-1)$, we have

$$\exp\left(- \frac{qt^2}{2c^2 n^{1/4}} \right) \leq \exp\left(- \frac{q\pi^2 n(2\ell-1)^2}{8a_j^2} \right)$$

and

$$\int_y^{y+\frac{\pi}{d_j}} e^{-\frac{1}{4}(t-m_j(t))^2} dt = \int_{-\frac{\pi}{2d_j}}^{\frac{\pi}{2d_j}} e^{-\frac{1}{4}(t-m_j(t))^2} dt$$

(8.71)
$$= \int_{-\frac{\pi}{2d_j}}^{\frac{\pi}{2d_j}} e^{-\frac{t^2}{4}} dt \leq 2\sqrt{\pi}.$$

for any y, since $\frac{1}{4}(t - m_j(t))^2$ is a periodic function. Thus, we get
(8.72)
$$\int_M^\infty \exp\left(-\frac{(t-m_j(t))^2}{4} - \frac{qt^2}{2c^2 n^{1/4}}\right)dt \leq \int_M^\infty e^{-\frac{t^2}{4}}dt + \sum_{\ell=1}^\infty 2\sqrt{\pi}\exp\left(-\frac{q\pi^2 n(2\ell-1)^2}{8a_j^2}\right).$$

Recall that $a_j \leq \sqrt{qn/2}$, hence
$$\sum_{\ell=1}^\infty \exp\left(-\frac{q\pi^2 n(2\ell-1)^2}{8a_j^2}\right) \leq \sum_{\ell=1}^\infty \exp\left(-\frac{\pi^2(2\ell-1)^2}{4}\right)$$
$$\leq \frac{e^{-\pi^2/4}}{1-e^{-\pi^2/2}} \leq \frac{1}{8}.$$

Plugging into (8.72), we get
$$\int_M^\infty \exp\left(-\frac{(t-m_j(t))^2}{4} - \frac{qt^2}{2c^2 n^{1/4}}\right)dt \leq 2\sqrt{\pi}\left(1 - \Phi\left(\frac{M}{\sqrt{2}}\right)\right) + \frac{\sqrt{\pi}}{4},$$
where $\Phi(\cdot)$ is the distribution function of $N(0,1)$. Plugging back into (8.69), we get

(8.73) $$\int_M^{\frac{\pi}{2} c n^{5/8}} |\phi_n(t)|dt \leq 2\sqrt{\pi}\left(1 - \Phi\left(\frac{M}{\sqrt{2}}\right)\right) + \frac{\sqrt{\pi}}{4}.$$

Now we go back to the first term of the right hand side of (8.66). Recall that $\phi_n(t)$ converge to $e^{-t^2/2}$ for all t. We have the following estimate:
$$\int_{-M}^M e^{-itx}\phi_n(t)dt$$
$$= \int_{-\infty}^\infty e^{-itx}e^{-t^2/2}dt - \left(\int_{-\infty}^{-M} + \int_M^\infty\right)e^{-itx}e^{-t^2/2}dt + \int_{-M}^M e^{-itx}(\phi_n(t) - e^{-t^2/2})dt$$
(8.74) $\geq \sqrt{2\pi}e^{-x^2/2} - 2\int_M^\infty e^{-t^2/2}dt - \int_{-M}^M |\phi_n(t) - e^{-t^2/2}|dt.$

Note that the second term of the left most side of (8.74) converges to 0 as $M \to \infty$ and for fixed M we have $\int_{-M}^M |\phi_n(t) - e^{-\frac{t^2}{2}}|dt \to 0$ by the Dominated Convergence Theorem. Plugging these and (8.73) into (8.66), we get
$$\liminf_{n \to \infty} \int_{-\frac{\pi}{2}cn^{5/8}}^{\frac{\pi}{2}cn^{5/8}} e^{-itx}\phi_n(t)dt \geq \sqrt{2\pi}e^{-x^2/2} - \frac{\sqrt{\pi}}{2},$$
which concludes the whole proof. $\square$

8.2. Starting at the $[0, n^{3/4}]$ regime: Proof of Theorem 8.2

THEOREM 8.23. *Let $I = [-An^{2/3}, An^{2/3}]$ where A is a fixed large constant. Then there exist positive constants K, a, q such that*

(8.75) $$\mathbb{P}\left(\tau_a \leq Kn^{1/4} \mid X_0 \in I\right) \geq q$$

where $\tau_a = \inf\{t \geq 0 : X_t \geq an^{3/4}\}$.

THEOREM 8.24. *For a constant A put $I = [-An^{2/3}, An^{2/3}]$ and $\tau = \inf\{t \geq 0 : X_t \in I\}$. Then there exist constant $c > 0$ such that for sufficiently large A, we have*
$$\mathbf{P}(\tau > t) \leq \frac{2|X_0|}{ct\sqrt{n}}.$$

Proof of Theorem 8.2: Let A, c be constants such that the assertion of Theorem 8.24 holds and write $I = [-An^{2/3}, An^{2/3}]$. Since $X_0 \leq n^{3/4}$, by Lemma 8.24 with $t = \frac{4n^{1/4}}{c}$, we have
$$\mathbb{P}(\tau \leq \frac{4n^{1/4}}{c}) > \frac{1}{2}$$
where $\tau = \min\{t : X_t \in I\}$. By Theorem 8.23 and the strong Markov property, we get that X_t exceeds $an^{3/4}$ within $(K + \frac{4}{c})n^{1/4}$ steps with probability at least $\frac{q}{2}$. By Theorem 8.1, we can couple X_t and with the stationary chain Y_t within $O(n^{1/4})$ steps such that they meet each other with probability $\Omega(1)$. Applying Lemma 3.2 concludes the proof. □

We now proceed to the proof of Theorem 8.23. We begin with some lemmas.

LEMMA 8.25. *Let A be a large constant and put $I = [-An^{2/3}, An^{2/3}]$. For any $q \in (0, 1)$, there exist a state $Z = Z(q) \in I$ and constants $a = a(q) > 0$, $K = K(q) > 0$ such that*

(8.76) $$\mathbf{P}\left(\tau_a \leq Kn^{1/4} | X_0 = Z\right) \geq q$$

where $\tau_a = \inf\{t \geq 0 : X_t \geq an^{3/4}\}$.

Proof. Let Y_t be a SW chain with $Y_0 \stackrel{d}{=} \pi_n$. By Theorem 8.6, there exists a constant B such that

(8.77) $$\pi_n([B^{-1}n^{3/4}, Bn^{3/4}]) \geq 1 - \frac{1-q}{4}.$$

We will prove the lemma for $a = B^{-1}$ and $K = \frac{6B}{c}$ where c is the constant in Theorem 8.24. Write $J = [B^{-1}n^{3/4}, Bn^{3/4}]$, then by (8.77) we have

(8.78) $$\mathbf{P}\left(Y_0 \in J \text{ and } Y_{Kn^{1/4}} \in J\right) \geq 1 - \frac{1-q}{2}.$$

Put $\tau = \inf\{t \geq 0 : Y_t \in I\}$. We have
$$\mathbf{P}(\tau \leq Kn^{1/4} | Y_0 \in J) \geq 1 - \frac{2Bn^{3/4}}{c\sqrt{n}Kn^{1/4}} = \frac{2}{3}$$
by Theorem 8.24. Thus we have
$$\mathbf{P}\left(Y_0 \in J, \tau \leq Kn^{1/4}\right) \geq \left(1 - \frac{1-q}{4}\right)\frac{2}{3} \geq \frac{1}{2}.$$

Let
$$\delta = \max_{W \in I} \mathbf{P}(\tau_a \leq Kn^{1/4} | X_0 = W).$$

By the strong Markov property
$$\mathbf{P}\left(Y_0 \in J, \tau \leq Kn^{1/4}, Y_{Kn^{1/4}} \leq an^{3/4}\right)$$
$$= \mathbf{P}\left(Y_0 \in J, \tau \leq Kn^{1/4}\right)\mathbf{P}\left(Y_{Kn^{1/4}} \leq an^{3/4} | \tau, Y_\tau\right) \geq \frac{1-\delta}{2},$$

since $\tau_a > Kn^{1/4}$ implies that $Y_{Kn^{1/4}} \leq an^{3/4}$. We deduce that
$$(8.79) \qquad \mathbf{P}\Big(Y_0 \in J, Y_{Kn^{1/4}} \notin J\Big) \geq \frac{1-\delta}{2}.$$
Combining (8.78) and (8.79) we get that $\delta \geq q$, concluding our proof. $\square$

LEMMA 8.26. *Consider the random graph* $G(\frac{n+X_0}{2}, \frac{2}{n})$ *where* $X_0 \in [-An^{2/3}, An^{2/3}]$ *for some large constant* A. *Then the intersection of the following events occurs with probability at least* $\delta = \delta(A) > 0$:
- $|\mathcal{C}_1| + |\mathcal{C}_2| \in [4An^{2/3}, 8An^{2/3}]$, *and* $|\mathcal{C}_2| > \frac{4A}{3}n^{2/3}$,
- $\mathcal{C}_1$ *and* $\mathcal{C}_2$ *are trees, and*
- $\sum_{j \geq 3} |\mathcal{C}_j|^2 \leq n^{4/3}$.

Proof. Let $\mathcal{A}$ be the event
- $|\mathcal{C}_1| + |\mathcal{C}_2| \in [4An^{2/3}, 8An^{2/3}], |\mathcal{C}_2| > \frac{4A}{3}n^{2/3}$,
- $\mathcal{C}_1$ and $\mathcal{C}_2$ are trees.

By Theorem 5.20 of [**16**], we have $\mathbf{P}(\mathcal{A}) \geq \delta = \delta(A) > 0$. Conditioned on $\mathcal{A}$ and on $\mathcal{C}_1$ and $\mathcal{C}_2$ the remaining graph, $\{\mathcal{C}_j\}_{j \geq 3}$, is distributed as $G(\frac{n+X_0}{2} - |\mathcal{C}_1| - |\mathcal{C}_2|, \frac{2}{n})$ conditioned to the event that it does not have components larger than $|\mathcal{C}_2|$. By Theorem 7 of [**25**] the complement of this event has probability decaying exponentially in A. Let $\{\mathcal{C}'_j\}$ be the component size in the unconditioned space $G(\frac{n+X_0}{2} - |\mathcal{C}_1| - |\mathcal{C}_2|, \frac{2}{n})$. We have
$$\mathbb{E} \sum_{j \geq 1} |\mathcal{C}'_j|^2 = \big(\frac{n+X_0}{2} - |\mathcal{C}_1| - |\mathcal{C}_2|\big) \mathbb{E} |\mathcal{C}(v)|.$$
Since $X_0 \leq An^{2/3}$ and $|\mathcal{C}_1| + |\mathcal{C}_2| \geq 4An^{2/3}$, Theorem 7 of [**25**] gives that
$$\mathbb{E} |\mathcal{C}(v)| \leq O(e^{-cA}) n^{1/3},$$
and so
$$\mathbb{E} \sum_{j \geq 1} |\mathcal{C}'_j|^2 \leq O(e^{-cA}) n^{4/3}.$$
The lemma now follows since in the conditioned space, the event we condition on has probability exponentially close to 1. $\square$

LEMMA 8.27. *Let* $I = [-An^{2/3}, An^{2/3}]$ *for some large* A. *There exist a constant* $c = c(A) > 0$ *such that*
$$(8.80) \qquad \mathbf{P}(X_1 = x \mid X_0 \in I) \geq cn^{-2/3}$$
for any $x \in n + 2\mathbb{Z}$ *with* $x \in I$ *and* $x > 0$.

Proof. Write $\mathcal{A}$ for the event of the assertion of Lemma 8.26 in $\{\mathcal{C}_j^+\}$, so that $\mathbf{P}(\mathcal{A}) \geq \delta(A) > 0$. In $G(\frac{n-|X_0|}{2}, \frac{2}{n})$ we have by Theorem 5.13 that
$$\mathbb{E} \sum_{j \geq 1} |\mathcal{C}_j^-|^2 \leq Dn^{4/3}$$
where $D = D(A)$ is a constant. We have by Markov's inequality that
$$\mathbf{P}\Big(|\sum_{j \geq 3} \epsilon_j |\mathcal{C}_j^+|| \leq Dn^{2/3} \Big| \mathcal{A}\Big) \geq 1 - 1/D^2,$$

and
$$\mathbf{P}(|\sum_{j\geq 1}\epsilon_j|\mathcal{C}_j^-|| \leq Dn^{2/3}) \geq 1 - 1/D^2,$$
and these two events are independent. Thus, the following event which we denote by $\mathcal{B}$ happens with probability $\Omega(1)$.
- $|\mathcal{C}_1^+| + |\mathcal{C}_2^+| \in [4An^{2/3}, 8An^{2/3}], |\mathcal{C}_2^+| > \frac{4A}{3}n^{2/3}$,
- $\mathcal{C}_1^+$ and $\mathcal{C}_2^+$ are trees,
- $|\sum_{j\geq 3}\epsilon_j|\mathcal{C}_j^+| + \sum_{j\geq 1}\epsilon_j|\mathcal{C}_j^-|| \leq 2Dn^{2/3}$.

Note that if a negative spin is assigned to $\mathcal{C}_2^+$ then
$$X_1 = |\mathcal{C}_1^+| - |\mathcal{C}_2^+| + \sum_{j\geq 3}\epsilon_j|\mathcal{C}_j^+| + \sum_{j\geq 1}\epsilon_j|\mathcal{C}_j^-|.$$

Thus
$$\mathbf{P}(X_1 = x) \geq \frac{1}{2}\mathbf{P}(|\mathcal{C}_1^+| - |\mathcal{C}_2^+| + \sum_{j\geq 3}\epsilon_j|\mathcal{C}_j^+| + \sum_{j\geq 1}\epsilon_j|\mathcal{C}_j^-| = x).$$

So we only need to show that for any $x \in [-An^{2/3}, An^{2/3}]$ we have
$$\mathbf{P}(|\mathcal{C}_1^+| - |\mathcal{C}_2^+| = x \mid \mathcal{B}) \geq cn^{-2/3}, \tag{8.81}$$
for some constant $c = c(A) > 0$. For any $m \in [4An^{2/3}, 8An^{2/3}]$ let $l = \frac{m+x}{2}$. By Cayley's formula we have that
$$\mathbf{P}(|\mathcal{C}_1^+| - |\mathcal{C}_2^+| = x \mid |\mathcal{C}_1^+| + |\mathcal{C}_2^+| = m, \mathcal{C}_1 \cup \mathcal{C}_2, \mathcal{B})$$
$$= \mathbf{P}(|\mathcal{C}_1^+| = l, |\mathcal{C}_2^+| = m - l \mid |\mathcal{C}_1^+| + |\mathcal{C}_2^+| = m, \mathcal{C}_1 \cup \mathcal{C}_2, \mathcal{B})$$
$$= \frac{\binom{m}{l}l^{l-2}(m-l)^{(m-l)-2}}{\sum_{k=\frac{4A}{3}n^{2/3}}^{m/2}\binom{m}{k}k^{k-2}(m-k)^{(m-k)-2}}. \tag{8.82}$$

Let
$$a(k) = \binom{m}{k}k^{k-2}(m-k)^{(m-k)-2}.$$

By Stirling's formula, there are two constants c and C such that for large enough n and any $k_1, k_2 \in [\frac{4A}{3}n^{2/3}, \frac{m}{2}]$, we have
$$c \leq \frac{a(k_1)}{a(k_2)} \leq C.$$

This implies
$$\frac{\binom{m}{k}k^{k-2}(m-k)^{(m-k)-2}}{\sum_{k=\frac{4A}{3}n^{2/3}}^{m/2}\binom{m}{k}k^{k-2}(m-k)^{(m-k)-2}} \geq cn^{-2/3}$$
which proves (8.81). $\square$

Proof of Theorem 8.23: Let A be large and $q \in (0,1)$ will be chosen later very close to 1. Let Z be the site and $K > 0$ the number satisfying the assertion of Lemma 8.25. Let $\{\widetilde{X}_t\}$ be an independent SW chain starting at Z and $\widetilde{\tau}_a$ is as in Lemma 8.25. Then we have
$$\mathbb{P}\left(\widetilde{\tau}_a \geq Kn^{1/4} \mid \widetilde{X}_0 = Z\right) \leq 1 - q. \tag{8.83}$$

Let $c > 0$ be the constant from Lemma 8.27. This lemma implies that we can couple X_t and $\widetilde{X}_t$ such that $X_1 = \widetilde{X}_1$ with probability at least c. From that point we can couple such that the two processes stay together with probability 1.

(8.84) $$\mathbb{P}(X_t = \widetilde{X}_t \text{ for } t \geq 1) \geq c.$$

Thus, we have

$$\mathbf{P}(\tau_a \leq Kn^{1/4}) \geq \mathbf{P}(X_t = \widetilde{X}_t \text{ for } t \geq 1 \text{ and } \widetilde{\tau}_a \leq Kn^{1/4}) \geq c - (1-q),$$

so we choose $q \geq 1 - c/2$ and conclude the proof. $\square$

To prove Theorem 8.24 we consider yet another modification of the SW dynamics $\{X'_t\}$. For any X'_0, in the supercritical random graph $G(\frac{n+|X_0|'}{2}, \frac{2}{n})$, let $\mathcal{C}_{\delta\epsilon n}$ be the component discovered by the exploration process at time $\delta\epsilon n$ where $\epsilon = \frac{X'_0}{n}$ and δ is a small constant (see Lemma 5.15). We assign positive spin to this component and random spins to all other components in $G(\frac{n+|X_0|'}{2}, \frac{2}{n})$ and all components in $G(\frac{n-|X_0|'}{2}, \frac{2}{n})$. Let X'_1 be the sum of spins after this assigning process.

The reason we require this change is that we were not able to obtain the bounds of Theorem 5.15 for $\mathcal{C}_1$, but only for $\mathcal{C}_{\delta\epsilon n}$ which is very likely to be $\mathcal{C}_1$. This will become evident in the proof. We first state a key lemma and then use it to prove Theorem 8.24.

LEMMA 8.28. *For any constant A put $I = [-An^{2/3}, An^{2/3}]$. Then there exists a constant $c > 0$ such that for sufficiently large A we have*

(8.85) $$\mathbb{E}\Big(|X'_1|\mathbf{1}_{\{X'_1 \notin I\}} + X'_1\mathbf{1}_{\{X'_1 \in I\}} \,\Big|\, |X'_0| > An^{2/3}\Big) < |X'_0| - c\sqrt{n}.$$

Proof of Theorem 8.24: Notice that of $|X_0| = |X'_0|$ then $|X_1| \stackrel{d}{=} |X'_1|$, and so $|X_t| \stackrel{d}{=} |X'_t|$ for all $t \geq 1$. Thus, we only need to prove the assertion of the Theorem for $\{X'_t\}$. For simplicity of notation we write X_t for X'_t. Assume that $|X_0| > An^{2/3}$ otherwise the assertion is trivial. We begin by noticing that
(8.86)
$$\mathbb{E}\Big(|X_{t+1}|\mathbf{1}_{\{\tau>t+1\}} + X_\tau \mathbf{1}_{\{\tau \leq t+1\}}\Big|\mathcal{F}_t\Big)\mathbf{1}_{\{\tau \leq t\}} = \mathbb{E}(X_\tau \mathbf{1}_{\{\tau \leq t\}}|\mathcal{F}_t) = X_\tau \mathbf{1}_{\{\tau \leq t\}}.$$

By Lemma 8.28 we have

$$\mathbb{E}\Big(|X_{t+1}|\mathbf{1}_{\{\tau>t+1\}} + X_\tau \mathbf{1}_{\{\tau \leq t+1\}}\Big|\mathcal{F}_t\Big)\mathbf{1}_{\{\tau \geq t+1\}} = \mathbb{E}\Big(|X_{t+1}|\mathbf{1}_{\{\tau>t+1\}} + X_\tau \mathbf{1}_{\{\tau = t+1\}}\Big|\mathcal{F}_t\Big)$$
(8.87) $$\leq |X_t|\mathbf{1}_{\{\tau>t\}} - c\sqrt{n}\mathbf{1}_{\{\tau>t\}}.$$

Thus, We have that $\{X_t\}$ satisfies the following inequality:
(8.88)
$$\mathbb{E}\Big(|X_{t+1}|\mathbf{1}_{\{\tau>t+1\}} + X_\tau \mathbf{1}_{\{\tau \leq t+1\}}\Big|\mathcal{F}_t\Big) \leq |X_t|\mathbf{1}_{\{\tau>t\}} + X_\tau \mathbf{1}_{\{\tau \leq t\}} - c\sqrt{n}\mathbf{1}_{\{\tau>t\}}.$$

Taking expectations of both sides of (8.88), we get

$$\mathbb{E}\Big(|X_{t+1}|\mathbf{1}_{\{\tau>t+1\}} + X_\tau \mathbf{1}_{\{\tau \leq t+1\}}\Big) \leq \mathbb{E}\Big(|X_t|\mathbf{1}_{\{\tau>t\}} + X_\tau \mathbf{1}_{\{\tau \leq t\}}\Big) - c\sqrt{n}\mathbb{P}(\tau > t).$$

Summing over t from 0 to $k-1$, we get

$$\mathbb{E}\Big(|X_k|\mathbf{1}_{\{\tau>k\}} + X_\tau\mathbf{1}_{\{\tau\leq k\}}\Big) \leq |X_0| - \sum_{t=0}^{k-1} c\sqrt{n}\mathbb{P}(\tau>t)$$

(8.89)
$$\leq |X_0| - kc\sqrt{n}\mathbb{P}(\tau>k).$$

We also have

$$\mathbb{E}\Big(|X_k|\mathbf{1}_{\{\tau>k\}} + X_\tau\mathbf{1}_{\{\tau\leq k\}}\Big) \geq \mathbb{E}\Big(X_\tau\mathbf{1}_{\{\tau\leq k\}}\Big) \geq -An^{2/3} \geq -|X_0|.$$

Combining this with (8.89), we have

$$kc\sqrt{n}\mathbb{P}(\tau>k) \leq 2|X_0|$$

which implies the required result. $\square$

LEMMA 8.29. *Let X be a random variable. Then for any $b < 0$ and positive integer k, we have*

$$\mathbb{E}\Big(|X|\mathbf{1}_{(X\leq b)}\Big) \leq \frac{\mathbb{E}|X|^k}{|b|^{k-1}}.$$

Proof of Lemma 8.29: We have

$$\mathbb{E}|X|^k \geq \int_{-\infty}^{b}(-x)^k dF(x) \geq |b|^{k-1}\int_{-\infty}^{b}(-x)dF(x) = |b|^{k-1}\mathbb{E}|X|\mathbf{1}_{(X\leq b)}.$$

$\square$

Proof of Lemma 8.28: For simplicity write again X_t for X'_t. Notice that

$$|X_1|\mathbf{1}_{\{X_1\notin I\}} + X_1\mathbf{1}_{\{X_1\in I\}} = X_1 + 2|X_1|\mathbf{1}_{\{X_1<-An^{2/3}\}}.$$

We first bound $\mathbb{E}X_1$ from above. Recall that in our modified chain we have

$$\mathbb{E}X_1 = \mathbb{E}|\mathcal{C}_{\delta\epsilon n}|.$$

By part (i) of Theorem 5.15, for sufficiently large A and $|X_0| \geq An^{2/3}$, we have

$$\mathbb{E}|\mathcal{C}_{\delta\epsilon n}| \leq 2\frac{X_0}{n}\frac{n+X_0}{2} - c\Big(\frac{X_0}{n}\Big)^{-2} = X_0 + \frac{X_0^2}{n} - \frac{cn^2}{X_0^2}.$$

If $X_0 \leq \sqrt[4]{\frac{c}{2}}n^{3/4}$, then we have $\frac{X_0^2}{n} \leq \frac{cn^2}{2X_0^2}$. In this case we have

(8.90)
$$\mathbb{E}X_1 = \mathbb{E}|\mathcal{C}_{\delta\epsilon n}| \leq X_0 - \frac{cn^2}{2X_0^2}.$$

If $X_0 > \sqrt[4]{\frac{c}{2}}n^{3/4}$, then by Lemma 8.16, we have

(8.91)
$$\mathbb{E}X_1 \leq X_0 - \frac{X_0^2}{6n}.$$

Next we bound $\mathbb{E}|X_1|\mathbf{1}_{\{X_1<-An^{2/3}\}}$ from above. Let

$$M = \sum_{\mathcal{C}_j^+ \neq \mathcal{C}_{\delta\epsilon n}^+} \epsilon_j|\mathcal{C}_j^+| + \sum_{j\geq 1}\epsilon'_j|\mathcal{C}_j^-|.$$

Then

$$X_1 = |\mathcal{C}_{\delta\epsilon n}| + M.$$

8. CRITICAL CASE

Since $|\mathcal{C}_{\delta\epsilon n}| > 0$, if $X_1 < -An^{2/3}$, then $M < -An^{2/3}$ and $|X_1| \leq -M$. Thus,
$$\mathbb{E}\Big(|X_1|\mathbf{1}_{\{X_1 < -An^{2/3}\}}\Big) \leq \mathbb{E}\Big((-M)\mathbf{1}_{\{X_1 < -An^{2/3}\}}\Big) \leq \mathbb{E}\Big((-M)\mathbf{1}_{\{M < -An^{2/3}\}}\Big).$$
Lemma 8.29 with $k = 4$ gives

(8.92) $$\mathbb{E}\Big((-M)\mathbf{1}_{\{M < -An^{2/3}\}}\Big) \leq \frac{\mathbb{E}M^4}{(An^{2/3})^3}.$$

We also have
$$\mathbb{E}M^4 \leq \sum_{\mathcal{C}_j^+ \neq \mathcal{C}_{\delta\epsilon n}^+} \mathbb{E}|\mathcal{C}_j^+|^4 + \sum_{j \geq 1} \mathbb{E}|\mathcal{C}_j^-|^4 + 6\Big[\sum_{\mathcal{C}_j^+ \neq \mathcal{C}_{\delta\epsilon n}^+} \mathbb{E}|\mathcal{C}_j^+|^2\Big]\Big[\sum_{j \geq 1} \mathbb{E}|\mathcal{C}_j^-|^2\Big]$$
(8.93) $$+ 6\Big[\sum_{\mathcal{C}_j^+, \mathcal{C}_i^+ \neq \mathcal{C}_{\delta\epsilon n}^+, i \neq j} \mathbb{E}|\mathcal{C}_i^+|^2|\mathcal{C}_j^+|^2 + \sum_{i,j \geq 1, i \neq j} \mathbb{E}|\mathcal{C}_i^-|^2|\mathcal{C}_j^-|^2\Big].$$

By (ii) of Theorem 5.15 and Lemma 5.12, we have

(8.94) $$\mathbb{E}M^4 = O\Big(\frac{n^6}{X_0^5}\Big) + O\Big(\frac{n^4}{X_0^2}\Big) = O\Big(\frac{n^4}{X_0^2}\Big)$$

since $|X_0| \geq An^{2/3}$. Plugging into (8.92), we have

(8.95) $$\mathbb{E}\Big(|X_1|\mathbf{1}_{\{X_1 < -An^{2/3}\}}\Big) = O\Big(\frac{n^2}{A^3 X_0^2}\Big).$$

If $X_0 \leq \sqrt[4]{\frac{c}{2}}n^{3/4}$, then combining (8.95) with (8.90) for large enough A, we get
$$\mathbb{E}X_1 + 2\mathbb{E}|X_1|\mathbf{1}_{\{X_1 < -An^{2/3}\}} \leq X_0 - \frac{cn^2}{4X_0^2} \leq X_0 - c\sqrt{n}.$$

If $X_0 > \sqrt[4]{\frac{c}{2}}n^{3/4}$, then combining (8.95) with (8.91) for large enough A, we get
$$\mathbb{E}X_1 + 2\mathbb{E}|X_1|\mathbf{1}_{\{X_1 < -An^{2/3}\}} \leq X_0 - \frac{X_0^2}{6n} + O(A^{-3}\frac{n^2}{X_0^2}) \leq X_0 - c\sqrt{n}.$$

Combining these two cases finishes the proof. □

8.3. The lower bound on the mixing time

Recall that in this section X_t is the original magnetization chain we defined in (4.2).

Proof of the lower bound of part (ii) of Theorem 2.1: Suppose X_t' is a modified magnetization chain and π' is its stationary distribution. By Theorem 8.6, we can choose an interval $[a_1 n^{3/4}, a_2 n^{3/4}]$ with $0 < a_1 < a_2$ such that

(8.96) $$\pi'(a_1 n^{3/4}, a_2 n^{3/4}) > \frac{3}{4}.$$

Suppose $X_0' = 3a_2 n^{3/4}$. By Theorem 8.18, there exists a constant k such that
$$\mathbf{P}(\tau > kn^{1/4}) \geq \frac{1}{2}$$
where τ is the first time that X_t exit $[a_2 n^{3/4}, 4a_2 n^{3/4}]$. This implies
$$\mathbf{P}\Big(X_{kn^{1/4}}' \geq a_2 n^{3/4}\Big) \geq \frac{1}{2}.$$

By Theorem 8.6, we have $\pi'(-a_2n^{3/4}, -a_1n^{3/4})$ converges to 0 as n goes to infinity. Also, by (8.9), we have $\mathbf{P}(X'_{kn^{1/4}} \in [-a_2n^{3/4}, -a_1n^{3/4}]) = O(n^{-1/12})$. Combining these, we get that

$$\pi'[(a_1n^{3/4}, a_2n^{3/4}), (-a_2n^{3/4}, -a_1n^{3/4})] - \mathbf{P}(X'_{kn^{1/4}} \in [(a_1n^{3/4}, a_2n^{3/4}), (-a_2n^{3/4}, -a_1n^{3/4})]) > \frac{1}{4}$$

for large enough n. Recall that $X_t \stackrel{d}{=} |X'_t|$, so this is equivalent to

$$\pi(a_1n^{3/4}, a_2n^{3/4}) - \mathbf{P}(X_{kn^{1/4}} \in (a_1n^{3/4}, a_2n^{3/4})) > \frac{1}{4},$$

i.e.,

(8.97) $$\left\| X_{kn^{1/4}} - \pi \right\|_{TV} > \frac{1}{4}.$$

CHAPTER 9

Fast mixing of the Swendsen-Wang process on trees

In this section we provide an upper bound estimate of the mixing time of the Swendsen-Wang process on any tree with n vertices. We will prove in a more general setting for The Swendsen-Wang process for the q-state ferromagnetic Potts model. Recall that Ising model is the case $q = 2$.

For any given graph $G = (V, E)$, consider the set $S = \{0,1\}^{|E|}$ of all edge configuration $\eta : E \to \{0, 1\}$. We consider the following Markov chain σ_t on S. At each step, we first color each component independently and uniformly from the q colors. Then we add all edges that connect vertices with the same color. Finally, delete each existing edge with probability $(1 - p)$ to get a new state in S. It is easy to see that this process the dual of the Swendsen-Wang process for the q-state ferromagnetic Potts model on vertices configurations and the stationary distribution of σ_t is the random cluster model. For any two Swendsen-Wang chains, if we can couple the corresponding edge models so that they are the same(i.e., they have same clusters) at some time, we therefore couple the original Swendsen-Wang process at the same time. Consequently, any upper bound of the mixing time of this edge model implies the same upper bound on ferromagnetic Potts model.

There is an exploration process on trees to present σ_t. Notice that on trees each edge with state 0 connects two separate components. For any given $\eta \in S$, we color each components independently and uniformly from the q colors, starting from the root. We add edges connects vertices with the same color. Notice that this procedure is equivalent to setting every edge originally has configuration 0 with configuration 1 with probability $\frac{1}{q}$ and maintain configuration 0 otherwise. Thus, the process σ_t can be described as follows: First change each edge of 0 to 1 with probability $\frac{1}{q}$ and stay 0 otherwise, independently for each of them. Then, change each edge of 1 ,including those who have changed from 0 to 1 in the previous step, to 0 with probability $1 - p$, and stay 1 otherwise, independently for each of them. Each bit evolves independently as a Markov chain on $\{0, 1\}$, with transition matrix

$$(9.1) \qquad \mathbf{p} = \begin{pmatrix} 1 - \frac{p}{Q} & \frac{p}{Q} \\ 1 - p & p \end{pmatrix}.$$

Proof of Theorem 2.2: The transition matrix (9.1) gives that we can couple every single edge with probability at least $1 - p + \frac{p}{q} \geq \frac{1}{q}$. Using the path coupling method of Bubley and Dyer (see Theorem 14.6 and Corollary 14.7 of [21]), we have

$$T_{\mathrm{mix}} \leq \frac{\log n + \log 4}{-\log p(1 - \frac{1}{q})}.$$

□

Acknowledgements

We are very grateful to Jian Ding for carefully reading the paper and for numerous useful suggestions. We also thank Lutz Warnke for pointing out some errors in a previous version of this paper.

Bibliography

[1] Noga Alon and Joel H. Spencer, *The probabilistic method*, 3rd ed., Wiley-Interscience Series in Discrete Mathematics and Optimization, John Wiley & Sons Inc., Hoboken, NJ, 2008. With an appendix on the life and work of Paul Erdős. MR2437651 (2009j:60004)

[2] Krishna B. Athreya and Peter E. Ney, *Branching processes*, Springer-Verlag, New York, 1972. Die Grundlehren der mathematischen Wissenschaften, Band 196. MR0373040 (51 #9242)

[3] Béla Bollobás, *The evolution of random graphs*, Trans. Amer. Math. Soc. **286** (1984), no. 1, 257–274, DOI 10.2307/1999405. MR756039 (85k:05090)

[4] Béla Bollobás and Oliver Riordan, *Asymptotic normality of the size of the giant component via a random walk*, J. Combin. Theory Ser. B **102** (2012), no. 1, 53–61, DOI 10.1016/j.jctb.2011.04.003. MR2871766 (2012m:05311)

[5] Christian Borgs, Jennifer T. Chayes, Alan Frieze, Jeong Han Kim, Prasad Tetali, Eric Vigoda, and Van Ha Vu, *Torpid mixing of some Monte Carlo Markov chain algorithms in statistical physics*, 40th Annual Symposium on Foundations of Computer Science (New York, 1999), IEEE Computer Soc., Los Alamitos, CA, 1999, pp. 218–229, DOI 10.1109/SFFCS.1999.814594. MR1917562

[6] Bubley R. and Dyer M. (1997), *Path Coupling: A technique for proving rapid mixing in Markov chains*. Proceeding of the 38th Annual Symposium FOCS (IEEE).

[7] Colin Cooper, Martin E. Dyer, Alan M. Frieze, and Rachel Rue, *Mixing properties of the Swendsen-Wang process on the complete graph and narrow grids*, J. Math. Phys. **41** (2000), no. 3, 1499–1527, DOI 10.1063/1.533194. Probabilistic techniques in equilibrium and nonequilibrium statistical physics. MR1757967 (2001g:82072)

[8] Colin Cooper and Alan M. Frieze, *Mixing properties of the Swendsen-Wang process on classes of graphs*, Random Structures Algorithms **15** (1999), no. 3-4, 242–261, DOI 10.1002/(SICI)1098-2418(199910/12)15:3/4¡242::AID-RSA4¿3.3.CO;2-3. Statistical physics methods in discrete probability, combinatorics, and theoretical computer science (Princeton, NJ, 1997). MR1716764 (2001b:60117)

[9] Burgess Davis and David McDonald, *An elementary proof of the local central limit theorem*, J. Theoret. Probab. **8** (1995), no. 3, 693–701, DOI 10.1007/BF02218051. MR1340834 (96j:60037)

[10] Richard Durrett, *Probability: theory and examples*, 2nd ed., Duxbury Press, Belmont, CA, 1996. MR1609153 (98m:60001)

[11] Robert G. Edwards and Alan D. Sokal, *Generalization of the Fortuin-Kasteleyn-Swendsen-Wang representation and Monte Carlo algorithm*, Phys. Rev. D (3) **38** (1988), no. 6, 2009–2012, DOI 10.1103/PhysRevD.38.2009. MR965465 (89i:82003)

[12] Richard S. Ellis, *Entropy, large deviations, and statistical mechanics*, Grundlehren der Mathematischen Wissenschaften [Fundamental Principles of Mathematical Sciences], vol. 271, Springer-Verlag, New York, 1985. MR793553 (87d:82008)

[13] P. Erdős and A. Rényi, *On the evolution of random graphs* (English, with Russian summary), Magyar Tud. Akad. Mat. Kutató Int. Közl. **5** (1960), 17–61. MR0125031 (23 #A2338)

[14] C. M. Fortuin and P. W. Kasteleyn, *On the random-cluster model. I. Introduction and relation to other models*, Physica **57** (1972), 536–564. MR0359655 (50 #12107)

[15] Vivek K. Gore and Mark R. Jerrum, *The Swendsen-Wang process does not always mix rapidly*, STOC '97 (El Paso, TX), ACM, New York, 1999, pp. 674–681 (electronic). MR1753371

[16] Svante Janson, Tomasz Łuczak, and Andrzej Rucinski, *Random graphs*, Wiley-Interscience Series in Discrete Mathematics and Optimization, Wiley-Interscience, New York, 2000. MR1782847 (2001k:05180)

[17] Mark Jerrum and Alistair Sinclair, *Polynomial-time approximation algorithms for the Ising model*, SIAM J. Comput. **22** (1993), no. 5, 1087–1116, DOI 10.1137/0222066. MR1237164 (94g:82007)

[18] Richard M. Karp, *The transitive closure of a random digraph*, Random Structures Algorithms **1** (1990), no. 1, 73–93, DOI 10.1002/rsa.3240010106. MR1068492 (91j:05093)

[19] Gregory F. Lawler and Vlada Limic, *Random walk: a modern introduction*, Cambridge Studies in Advanced Mathematics, vol. 123, Cambridge University Press, Cambridge, 2010. MR2677157 (2012a:60132)

[20] David A. Levin, Malwina J. Luczak, and Yuval Peres, *Glauber dynamics for the mean-field Ising model: cut-off, critical power law, and metastability*, Probab. Theory Related Fields **146** (2010), no. 1-2, 223–265, DOI 10.1007/s00440-008-0189-z. MR2550363 (2011d:82062)

[21] David A. Levin, Yuval Peres, and Elizabeth L. Wilmer, *Markov chains and mixing times*, American Mathematical Society, Providence, RI, 2009. With a chapter by James G. Propp and David B. Wilson. MR2466937 (2010c:60209)

[22] Tomasz Luczak, *Component behavior near the critical point of the random graph process*, Random Structures Algorithms **1** (1990), no. 3, 287–310, DOI 10.1002/rsa.3240010305. MR1099794 (92c:05139)

[23] Anders Martin-Löf, *Symmetric sampling procedures, general epidemic processes and their threshold limit theorems*, J. Appl. Probab. **23** (1986), no. 2, 265–282. MR839984 (87i:92039)

[24] Asaf Nachmias and Yuval Peres, *Component sizes of the random graph outside the scaling window*, ALEA Lat. Am. J. Probab. Math. Stat. **3** (2007), 133–142. MR2349805 (2008i:05174)

[25] Asaf Nachmias and Yuval Peres, *The critical random graph, with martingales*, Israel J. Math. **176** (2010), 29–41, DOI 10.1007/s11856-010-0019-8. MR2653185 (2011i:05225)

[26] Asaf Nachmias and Yuval Peres, *Critical percolation on random regular graphs*, Random Structures Algorithms **36** (2010), no. 2, 111–148, DOI 10.1002/rsa.20277. MR2583058 (2011f:60195)

[27] Persky N., Ben-Av R., Kanter I. and Domany E. (1996), Mean-field behavior of cluster dynamics, *Phys. Rev. E* **54**, 2351-2358.

[28] Boris Pittel, *On tree census and the giant component in sparse random graphs*, Random Structures Algorithms **1** (1990), no. 3, 311–342, DOI 10.1002/rsa.3240010306. MR1099795 (92f:05087)

[29] Boris Pittel and Nicholas C. Wormald, *Counting connected graphs inside-out*, J. Combin. Theory Ser. B **93** (2005), no. 2, 127–172, DOI 10.1016/j.jctb.2004.09.005. MR2117934 (2005m:05117)

[30] Ray T., Tamayo P. and Klein W. (1989), Mean-field study of the Swendsen-Wang dynamics. *Phys. Rev. A* **39**, 5949-5953.

[31] J. Salas, *Dynamic critical behavior of cluster algorithms for 2D Ashkin-Teller and Potts models*, Markov Process. Related Fields **7** (2001), no. 1, 55–74. Inhomogeneous random systems (Cergy-Pontoise, 2000). MR1835747

[32] Jesús Salas and Alan D. Sokal, *Dynamic critical behavior of a Swendsen-Wang-type algorithm for the Ashkin-Teller model*, J. Statist. Phys. **85** (1996), no. 3-4, 297–361, DOI 10.1007/BF02174209. MR1413666 (98a:82065)

[33] Salas J. and Sokal A. D. (1997), Dynamic critical behavior of the Swendsen-Wang algorithm: the two-dimensional three-state Potts model revisited. *J. Stat. Phys.* **87** 1-36.

[34] Frank Spitzer, *A combinatorial lemma and its application to probability theory*, Trans. Amer. Math. Soc. **82** (1956), 323–339. MR0079851 (18,156e)

[35] Swendsen R. H. and Wang J. S. (1987), Nonuniversal critical dynamics in Monte Carlo simulations. Phys. Rev. Lett. **58**, 86-88.

[36] Wang J. S. (1990), Critical dynamics of the Swendsen-Wang algorithm in the three-dimensional Ising model. *Physica A* **164**, 240-244.

Editorial Information

To be published in the *Memoirs*, a paper must be correct, new, nontrivial, and significant. Further, it must be well written and of interest to a substantial number of mathematicians. Piecemeal results, such as an inconclusive step toward an unproved major theorem or a minor variation on a known result, are in general not acceptable for publication.

Papers appearing in *Memoirs* are generally at least 80 and not more than 200 published pages in length. Papers less than 80 or more than 200 published pages require the approval of the Managing Editor of the Transactions/Memoirs Editorial Board. Published pages are the same size as those generated in the style files provided for $\mathcal{AMS}$-LaTeX or $\mathcal{AMS}$-TeX.

Information on the backlog for this journal can be found on the AMS website starting from http://www.ams.org/memo.

A Consent to Publish is required before we can begin processing your paper. After a paper is accepted for publication, the Providence office will send a Consent to Publish and Copyright Agreement to all authors of the paper. By submitting a paper to the *Memoirs*, authors certify that the results have not been submitted to nor are they under consideration for publication by another journal, conference proceedings, or similar publication.

Information for Authors

Memoirs is an author-prepared publication. Once formatted for print and on-line publication, articles will be published as is with the addition of AMS-prepared frontmatter and backmatter. Articles are not copyedited; however, confirmation copy will be sent to the authors.

Initial submission. The AMS uses Centralized Manuscript Processing for initial submissions. Authors should submit a PDF file using the Initial Manuscript Submission form found at www.ams.org/submission/memo, or send one copy of the manuscript to the following address: Centralized Manuscript Processing, MEMOIRS OF THE AMS, 201 Charles Street, Providence, RI 02904-2294 USA. If a paper copy is being forwarded to the AMS, indicate that it is for *Memoirs* and include the name of the corresponding author, contact information such as email address or mailing address, and the name of an appropriate Editor to review the paper (see the list of Editors below).

The paper must contain a *descriptive title* and an *abstract* that summarizes the article in language suitable for workers in the general field (algebra, analysis, etc.). The *descriptive title* should be short, but informative; useless or vague phrases such as "some remarks about" or "concerning" should be avoided. The *abstract* should be at least one complete sentence, and at most 300 words. Included with the footnotes to the paper should be the 2010 *Mathematics Subject Classification* representing the primary and secondary subjects of the article. The classifications are accessible from www.ams.org/msc/. The Mathematics Subject Classification footnote may be followed by a list of *key words and phrases* describing the subject matter of the article and taken from it. Journal abbreviations used in bibliographies are listed in the latest *Mathematical Reviews* annual index. The series abbreviations are also accessible from www.ams.org/msnhtml/serials.pdf. To help in preparing and verifying references, the AMS offers MR Lookup, a Reference Tool for Linking, at www.ams.org/mrlookup/.

Electronically prepared manuscripts. The AMS encourages electronically prepared manuscripts, with a strong preference for $\mathcal{AMS}$-LaTeX. To this end, the Society has prepared $\mathcal{AMS}$-LaTeX author packages for each AMS publication. Author packages include instructions for preparing electronic manuscripts, samples, and a style file that generates the particular design specifications of that publication series. Though $\mathcal{AMS}$-LaTeX is the highly preferred format of TeX, author packages are also available in $\mathcal{AMS}$-TeX.

Authors may retrieve an author package for *Memoirs of the AMS* from www.ams.org/journals/memo/memoauthorpac.html or via FTP to ftp.ams.org (login as anonymous, enter your complete email address as password, and type cd pub/author-info). The

AMS Author Handbook and the *Instruction Manual* are available in PDF format from the author package link. The author package can also be obtained free of charge by sending email to `tech-support@ams.org` or from the Publication Division, American Mathematical Society, 201 Charles St., Providence, RI 02904-2294, USA. When requesting an author package, please specify $\mathcal{AMS}$-LaTeX or $\mathcal{AMS}$-TeX and the publication in which your paper will appear. Please be sure to include your complete mailing address.

After acceptance. The source files for the final version of the electronic manuscript should be sent to the Providence office immediately after the paper has been accepted for publication. The author should also submit a PDF of the final version of the paper to the editor, who will forward a copy to the Providence office.

Accepted electronically prepared files can be submitted via the web at `www.ams.org/submit-book-journal/`, sent via FTP, or sent on CD to the Electronic Prepress Department, American Mathematical Society, 201 Charles Street, Providence, RI 02904-2294 USA. TeX source files and graphic files can be transferred over the Internet by FTP to the Internet node `ftp.ams.org` (130.44.1.100). When sending a manuscript electronically via CD, please be sure to include a message indicating that the paper is for the *Memoirs*.

Electronic graphics. Comprehensive instructions on preparing graphics are available at `www.ams.org/authors/journals.html`. A few of the major requirements are given here.

Submit files for graphics as EPS (Encapsulated PostScript) files. This includes graphics originated via a graphics application as well as scanned photographs or other computer-generated images. If this is not possible, TIFF files are acceptable as long as they can be opened in Adobe Photoshop or Illustrator.

Authors using graphics packages for the creation of electronic art should also avoid the use of any lines thinner than 0.5 points in width. Many graphics packages allow the user to specify a "hairline" for a very thin line. Hairlines often look acceptable when proofed on a typical laser printer. However, when produced on a high-resolution laser imagesetter, hairlines become nearly invisible and will be lost entirely in the final printing process.

Screens should be set to values between 15% and 85%. Screens which fall outside of this range are too light or too dark to print correctly. Variations of screens within a graphic should be no less than 10%.

Inquiries. Any inquiries concerning a paper that has been accepted for publication should be sent to `memo-query@ams.org` or directly to the Electronic Prepress Department, American Mathematical Society, 201 Charles St., Providence, RI 02904-2294 USA.

Editors

This journal is designed particularly for long research papers, normally at least 80 pages in length, and groups of cognate papers in pure and applied mathematics. Papers intended for publication in the *Memoirs* should be addressed to one of the following editors. The AMS uses Centralized Manuscript Processing for initial submissions to AMS journals. Authors should follow instructions listed on the Initial Submission page found at www.ams.org/memo/memosubmit.html.

Algebra, to ALEXANDER KLESHCHEV, Department of Mathematics, University of Oregon, Eugene, OR 97403-1222; e-mail: klesh@uoregon.edu

Algebraic geometry, to DAN ABRAMOVICH, Department of Mathematics, Brown University, Box 1917, Providence, RI 02912; e-mail: amsedit@math.brown.edu

Algebraic topology, to SOREN GALATIUS, Department of Mathematics, Stanford University, Stanford, CA 94305 USA; e-mail: transactions@lists.stanford.edu

Arithmetic geometry, to TED CHINBURG, Department of Mathematics, University of Pennsylvania, Philadelphia, PA 19104-6395; e-mail: math-tams@math.upenn.edu

Automorphic forms, representation theory and combinatorics, to DANIEL BUMP, Department of Mathematics, Stanford University, Building 380, Sloan Hall, Stanford, California 94305; e-mail: bump@math.stanford.edu

Combinatorics and discrete geometry, to IGOR PAK, Department of Mathematics, University of California, Los Angeles, California 90095; e-mail: pak@math.ucla.edu

Commutative and homological algebra, to LUCHEZAR L. AVRAMOV, Department of Mathematics, University of Nebraska, Lincoln, NE 68588-0130; e-mail: avramov@math.unl.edu

Differential geometry and global analysis, to CHRIS WOODWARD, Department of Mathematics, Rutgers University, 110 Frelinghuysen Road, Piscataway, NJ 08854; e-mail: ctw@math.rutgers.edu

Dynamical systems and ergodic theory and complex analysis, to YUNPING JIANG, Department of Mathematics, CUNY Queens College and Graduate Center, 65-30 Kissena Blvd., Flushing, NY 11367; e-mail: Yunping.Jiang@qc.cuny.edu

Ergodic theory and combinatorics, to VITALY BERGELSON, Ohio State University, Department of Mathematics, 231 W. 18th Ave, Columbus, OH 43210; e-mail: vitaly@math.ohio-state.edu

Functional analysis and operator algebras, to STEFAAN VAES, KU Leuven, Department of Mathematics, Celestijnenlaan 200B, B-3001 Leuven, Belgium; e-mail: stefaan.vaes@wis.kuleuven.be

Geometric analysis, to WILLIAM P. MINICOZZI II, Department of Mathematics, Johns Hopkins University, 3400 N. Charles St., Baltimore, MD 21218; e-mail: trans@math.jhu.edu

Geometric topology, to MARK FEIGHN, Math Department, Rutgers University, Newark, NJ 07102; e-mail: feighn@andromeda.rutgers.edu

Harmonic analysis, complex analysis, to MALABIKA PRAMANIK, Department of Mathematics, 1984 Mathematics Road, University of British Columbia, Vancouver, BC, Canada V6T 1Z2; e-mail: malabika@math.ubc.ca

Harmonic analysis, representation theory, and Lie theory, to E. P. VAN DEN BAN, Department of Mathematics, Utrecht University, P.O. Box 80 010, 3508 TA Utrecht, The Netherlands; e-mail: E.P.vandenBan@uu.nl

Logic, to ANTONIO MONTALBAN, Department of Mathematics, The University of California, Berkeley, Evans Hall #3840, Berkeley, California, CA 94720; e-mail: antonio@math.berkeley.edu

Number theory, to SHANKAR SEN, Department of Mathematics, 505 Malott Hall, Cornell University, Ithaca, NY 14853; e-mail: ss70@cornell.edu

Partial differential equations, to MARKUS KEEL, School of Mathematics, University of Minnesota, Minneapolis, MN 55455; e-mail: keel@math.umn.edu

Partial differential equations and functional analysis, to ALEXANDER KISELEV, Department of Mathematics, University of Wisconsin-Madison, 480 Lincoln Dr., Madison, WI 53706; e-mail: kisilev@math.wisc.edu

Probability and statistics, to PATRICK FITZSIMMONS, Department of Mathematics, University of California, San Diego, 9500 Gilman Drive, La Jolla, CA 92093-0112; e-mail: pfitzsim@math.ucsd.edu

Real analysis and partial differential equations, to WILHELM SCHLAG, Department of Mathematics, The University of Chicago, 5734 South University Avenue, Chicago, IL 60615; e-mail: schlag@math.uchicago.edu

All other communications to the editors, should be addressed to the Managing Editor, ALEJANDRO ADEM, Department of Mathematics, The University of British Columbia, Room 121, 1984 Mathematics Road, Vancouver, B.C., Canada V6T 1Z2; e-mail: adem@math.ubc.ca

Selected Published Titles in This Series

1088 Mark Green, Phillip Griffiths, and Matt Kerr, Special Values of Automorphic Cohomology Classes, 2014

1087 Colin J. Bushnell and Guy Henniart, To an Effective Local Langlands Correspondence, 2014

1086 Stefan Ivanov, Ivan Minchev, and Dimiter Vassilev, Quaternionic Contact Einstein Structures and the Quaternionic Contact Yamabe Problem, 2014

1085 A. L. Carey, V. Gayral, A. Rennie, and F. A. Sukochev, Index Theory for Locally Compact Noncommutative Geometries, 2014

1084 Michael S. Weiss and Bruce E. Williams, Automorphisms of Manifolds and Algebraic K-Theory: Part III, 2014

1083 Jakob Wachsmuth and Stefan Teufel, Effective Hamiltonians for Constrained Quantum Systems, 2014

1082 Fabian Ziltener, A Quantum Kirwan Map: Bubbling and Fredholm Theory for Symplectic Vortices over the Plane, 2014

1081 Sy-David Friedman, Tapani Hyttinen, and Vadim Kulikov, Generalized Descriptive Set Theory and Classification Theory, 2014

1080 Vin de Silva, Joel W. Robbin, and Dietmar A. Salamon, Combinatorial Floer Homology, 2014

1079 Pascal Lambrechts and Ismar Volić, Formality of the Little N-disks Operad, 2013

1078 Milen Yakimov, On the Spectra of Quantum Groups, 2013

1077 Christopher P. Bendel, Daniel K. Nakano, Brian J. Parshall, and Cornelius Pillen, Cohomology for Quantum Groups via the Geometry of the Nullcone, 2013

1076 Jaeyoung Byeon and Kazunaga Tanaka, Semiclassical Standing Waves with Clustering Peaks for Nonlinear Schrödinger Equations, 2013

1075 Deguang Han, David R. Larson, Bei Liu, and Rui Liu, Operator-Valued Measures, Dilations, and the Theory of Frames, 2013

1074 David Dos Santos Ferreira and Wolfgang Staubach, Global and Local Regularity of Fourier Integral Operators on Weighted and Unweighted Spaces, 2013

1073 Hajime Koba, Nonlinear Stability of Ekman Boundary Layers in Rotating Stratified Fluids, 2014

1072 Victor Reiner, Franco Saliola, and Volkmar Welker, Spectra of Symmetrized Shuffling Operators, 2014

1071 Florin Diacu, Relative Equilibria in the 3-Dimensional Curved n-Body Problem, 2014

1070 Alejandro D. de Acosta and Peter Ney, Large Deviations for Additive Functionals of Markov Chains, 2014

1069 Ioan Bejenaru and Daniel Tataru, Near Soliton Evolution for Equivariant Schrödinger Maps in Two Spatial Dimensions, 2014

1068 Florica C. Cîrstea, A Complete Classification of the Isolated Singularities for Nonlinear Elliptic Equations with Inverse Square Potentials, 2014

1067 A. González-Enríquez, A. Haro, and R. de la Llave, Singularity Theory for Non-Twist KAM Tori, 2014

1066 José Ángel Peláez and Jouni Rättyä, Weighted Bergman Spaces Induced by Rapidly Increasing Weights, 2014

1065 Emmanuel Schertzer, Rongfeng Sun, and Jan M. Swart, Stochastic Flows in the Brownian Web and Net, 2014

1064 J. L. Flores, J. Herrera, and M. Sánchez, Gromov, Cauchy and Causal Boundaries for Riemannian, Finslerian and Lorentzian Manifolds, 2013

For a complete list of titles in this series, visit the
AMS Bookstore at **www.ams.org/bookstore/memoseries/**.